I

II

THE PLANT
BETWEEN SUN AND EARTH

and the Science of Physical and Ethereal Spaces

GEORGE ADAMS, *M.A.Cantab.* / OLIVE WHICHER

With 20 colour plates and
102 illustrations in black and white

PREFACE BY

EHRENFRIED PFEIFFER, M.D.(HON.)

RUDOLF STEINER PRESS, LONDON

First Edition, Goethean Science Foundation, Clent 1952
Second Edition (revised and enlarged by Olive Whicher) Rudolf Steiner Press,
    London 1980
© Rudolf Steiner Press, London 1980

Colour Plates
© Verlag Freies Geistesleben, Stuttgart 1979
Printed by Greiserdruck, Rastatt

ISBN 0 85440 360 4
Cover design by John Playfoot and Michael Wilson

Made and printed in Great Britain at
The Camelot Press Ltd, Southampton

PREFACE

EHRENFRIED PFEIFFER

The Universe in which we live is organized according to the principles of measure, number, and balanced, harmonious relation. Natural laws follow a pattern which can be re-discovered by human science. Geometry and mathematics play an important part in the evaluation of natural laws. Human intelligence and reasoning discern the rules and principles according to which the Universe and the kingdoms of Nature are built. It is the cosmic intelligence of the Creator which has built and formed the Cosmos. The best that man can do, using his own intelligent powers, is to describe how and by what means the Creator has formed the world.

The philosopher Schelling once said: 'To know Nature is, in effect, to re-create the world in man's own mind.' This is true science. Man followed in the Creator's footsteps, learning to know the creative thought and deed. He stands in awe and admiration before a superior intelligence, acknowledging a master mind. The total achievement of science hitherto is but a recognition of a small fraction of the creative thoughts by which the world is made. Matter with its complicated structures, functions and relations in the inorganic and in the living world is one side of the picture; the other side is Form. To recognize the formative forces is no less important than to gain knowledge of the functions and functional relations of matter as such. It is the formative forces which in the living world organize the matter. What we behold in the form and organic structure of any and every plant is a direct expression of the cosmic and creative thinking.

When science tells of the very intricate chemical and physiological nature of plants, it is pointing to a realm of harmonious relations, describable in terms of number and proportion – such as the constant and multiple proportions of chemical reactions. But the organizing factor in the realm of growth and life, in all living organs, is also describable in scientific terms. This organizing factor assigns to each organ of the plant – to leaf and root, fruit and flower – its specific form. Fascinated during recent decades by the intricate relations inherent in the structure of matter, scientists have neglected the study and recognition of those forces which arrange matter into the shape and pattern of living things. It is the merit of this book to have applied geometrical imagination and mathematical reasoning in recognizing the formative forces which can be seen at work in living Nature, thus discovering – or re-creating, in Schelling's sense – a further realm in which the cosmic intelligence is manifested.

3

Goethe's organic concept and Rudolf Steiner's imaginative idea of the formative forces have inspired the authors in this work, opening the way to a scientific understanding of the principles by which the living world is formed. The reader finds himself at the threshold of a new world; he begins to see the organizing principles in Nature as an expression of realities which heretofore were acknowledged only by the artistic and aesthetic sense. These now become accessible to science.

It was necessary, in order to describe these principles of form, to follow the path of the essentially new ideas which are contained in modern Projective Geometry; in fact, to develop a quite new geometry, namely that of living things. The very concept of space is enlarged, for the organic form is shown to contain more than the mere outward three-dimensional space. Without ever having to become too abstract or being lost in technical details, the reader is guided so that he penetrates right through this outward space into that transcendent space-dimension which exists in the Creator's mind. Whether we name it 'spiritual space', or the 'etheric' world, or even a 'fourth dimension' – terms do not matter – we begin to see with the eyes of a higher level of intelligence. A bridge is built between art and science, leading to a growing interpenetration of the aesthetic sense with knowledge such as a scientific mind will require.

It speaks for the authors' painstaking study and endeavour that they have here been able to transmit their own vision and research in a way that every one – layman as well as highly trained searcher for the truth – can follow without difficulty. The illustrations convey in imaginative form what geometrical construction and reasoning explain to the thinking mind. The two together form another step towards a recognition of the Master Architect's plan.

In a Preface there is no need to go into the documentation or other details of the book. It is sufficient to point out that we are here confronted with an unique and original approach, worthy of serious consideration by all who desire to enlarge their horizon and reach beyond the mere materialistic concept into a realm of creative, spiritual knowledge. The materialist can follow this method without losing what is essential to his point of view – the need for exact, mathematically valid reasoning. At the same time he is safely guided to the threshold of things to come: a new science of organic form and pattern.

To anyone already familiar with the reality of a spiritual conception of the world, this book will be a welcome contribution to the study of Nature, guiding him from imagination into the realm of the visible manifestation of the Spirit. In the historic conversation between Goethe and Schiller on the Archetypal Plant or 'Ur-Plant' – the principle according to which all plants are constructed – Schiller objected that this was only an 'idea'. Goethe replied, 'Then I can only be glad that I have ideas which I can even see with my eyes.' Goethe was asking for the development of what he called *anschauende Urteilskraft* or 'perceptive judgement' – judgement, recognition of the truth, in the very act of seeing. The present work, *The Plant between Sun and Earth*, is an exercise in this 'judgement in the act of seeing', and is thus an important step towards a truly Goethean science.

CONTENTS

Chapter VIII *The World of the Flower – Fulfilment*

FIGURES IN BLACK AND WHITE

8

9

ACKNOWLEDGEMENTS

Most of the plant drawings are from Nature. We acknowledge help derived from *Traité Général de Botanique* by Maout and Decaisne (1876) for the *Basella alba* (Plate XVI), *Aquilegia* in Figure 91 and some of the detail in Figure 89. The Brazil Nut (Figure 18) is adapted, by kind permission of the publishers, the C. W. Daniel Company Ltd., from an illustration by Margaret Fuller in *Fruit and Flowers* by Constance Garlick (1924). The Tulip (Figure 93) and the Proliferous Rose (Figure 94) are based on illustrations, reproduced in J. Schuster's illustrated edition of the *Metamorphosis of Plants* (1924) and attributed, partly at least, to Goethe's own hand. For the completion of some of the geometrical figures in black and white, we are indebted to the late Louis Loynes.

The illustrations of microscopic buds in Figure 58, photographed by Professor Ottilie Zeller, are included with her permission, and for the microscopic photographs in Figures 59, 66 and 97 we are indebted to Michael Wisniewski. Figures 78, 79, 80 and 101 are due to Schwenk, 81 to Gallerie Wilde, 84 and 85 to Grohmann, 99 to Selawry, 100 to Fyfe, and 102 to the Pfeiffer Foundation. Figure 90 is from Hortus Eystettensis.

INTRODUCTION

To the First Edition

In the Goethe Bicentenary year 1949 a new scientific theory was put forward by the authors of the present work, concerning the formative forces of the living world. The leading ideas of Goethe's *Metamorphosis of Plants* were interpreted and developed in a fresh direction, with the help of scientific methods not yet available in Goethe's time.

An essentially new step in science often has to cut across long-established boundaries between special fields, and so it is in this instance. While the main subject-matter is botanical, the methods brought to bear on the phenomena of plant form are derived from the modern projective school of geometry. Though not abstruse or recondite in its essential nature, this geometry is far too little known, and if expressed in the conventional scientific form the new morphological ideas would have been intelligible only to mathematically trained readers. We, therefore, had recourse to a more pictorial and imaginative form of presentation than is usual in scientific work. The sequence of thought and its connection with the phenomena of plant life was set forth in a large number of coloured drawings, exhibited in connection with the Bicentenary celebrations. We published at the same time a monograph entitled: *The Living Plant and the Science of Physical and Ethereal Spaces* (Goethean Science Foundation, 1949).

A selection of the drawings is reproduced for the first time in the present work, in which the whole conception is once more put forward. In Part I, the phenomena are explained in direct connection with the pictures, which, with the help of comparatively few words, are allowed to speak for themselves. Nature contains her own evidence of fundamental structure. A theory, if true, will be the outcome of a healthy perception of phenomena and will appeal to the sense of reality even without full scientific explanation. In Part II the fundamental theory is set forth more explicitly. Parts I and II, with the illustrations, should provide layman and scientist alike with a fair introduction to the idea of "ethereal spaces" and formative processes as applied in particular to the world of plants. (Technical details are briefly dealt with in the Notes, where other relevant works are also mentioned. In an Appendix the illustrations are correlated with the aforesaid monograph *The Living Plant*, to which the reader may refer for further study.) [This paragraph refers to the form of the first edition; the present second edition adheres to that of the (later) German edition. Ed.]

The phenomena of life show an expansive tendency of upward and outward growth very different in character from the tendencies of inert matter. In seeking to understand natural phenomena, science was dominated – until the end of the nineteenth century – by the idea of forces proceeding like gravity or the electric and magnetic attractions and repulsions from one point-centre to another. This was a natural outcome of the geometry of Euclid, which had been deeply rooted in the minds of men for two thousand years or more when in the age of Galileo, Descartes and Newton the modern scientific study of the world began. Meanwhile the new projective school of geometrical thinking was in its first beginnings, and when at last during the 19th century it blossomed forth, it remained "pure mathematics". With few exceptions, the significance for natural science of these quite new ideas concerning space and spatial form was lost amid the epoch-making discoveries of the time, in which the concept of a world based entirely on the centric or gravitational type of force went on from triumph to triumph.

Today it is widely recognized that science has reached the threshold of a new era. The most firmly established theories are being modified; remaining relatively valid, they gain quite another aspect when seen in a wider setting. Another far-reaching change of this kind is here suggested. From the essential polarity of spatial structure revealed in Projective Geometry (the so-called "Principle of Duality") is derived a scientific concept of formative forces specific to the living world, manifested in the phenomena of growth and form. These forces are not centric but peripheral in nature; moreover in tendency they are anti-gravitational. In contrast to the "physical", we therefore describe them as "ethereal" forces, for in this way the concept of the ethereal can once again be given clear definition. In a living body there is always the polar interplay between the two. The plant is a physical form imbued with life – a synthesis of the physical with the ethereal.

The new conception gives a way of access to the essential wholeness of living things, complementing the one-sided approach which seeks the source of life in ever smaller particles of the material substance. It is here applied to plant growth and form as seen in living Nature. At this initial stage no attempt is made to establish the link with the prevailing notions of the atomic and electronic realm. Nor has the concept yet been followed into the cytological domain.

In the Plates which accompany this volume, colour is used not arbitrarily but according to its inherent law, expressed in Goethe's *Theory of Colour*. For example, a sequence of colours will indicate the mathematical sequence in a family of curves. Complementary colours, notably green and a rosy pink – "peach-blossom", Goethe called it – are used to demonstrate polarities. Colour is also used to convey the idea of different kinds and qualities of space. Where it is not only a question of the finished form but of the spatial field from which the form emerges, this makes it possible to evoke in imagination the same truths which would otherwise be expressed in mathematical terms.

Since the publication of *The Living Plant* three years ago, a number of works have appeared, relevant to the present theme. Dr. Agnes Arber's *The Natural*

Philosophy of Plant Form (Cambridge University Press, 1950) contains an appreciation of Goethe's botanical work, probably the most important that has yet appeared by any English writer. Dr. E. Lehrs in *Man or Matter* (Faber and Faber, 1951) develops in a wider field the concept of peripheral and centric forces, making many detailed suggestions which deserve consideration. Reviewing the whole period of scientific consciousness from Galileo until the present time, he indicates how and why it is that the peripheral forces have remained hitherto unrecognized. Finally we would mention a work which from quite other points of view makes the bold suggestion that the source of life is in the wider Universe and that the very substance of the Earth is constantly renewed by the living process. We refer to W. Branfield's *Continuous Creation* (Routledge, 1950).

The authors acknowledge their great indebtedness to Rudolf Steiner, a pioneer among the modern interpreters of Goethe's scientific method, who also showed how it might in our time be renewed and carried further. Already in the 1880's, Rudolf Steiner was deriving from the theory of knowledge implicit in Goethe's work the philosophic justification for a more imaginative method. He was convinced that the scientific age would soon be reaching forward to far deeper problems, concerned above all with the realms of life and consciousness. Rudolf Steiner found a new approach to those more cosmic formative forces of the living world of which old traditions tell, and which he too described as the "ethereal". He recognized their character as peripheral or planar forces, and spoke of them as working surface-wise from the wide spaces of the Universe towards the living entities on Earth. He pointed to the importance of Projective Geometry, both in method and in content, for a future science of living Nature.

GEORGE ADAMS
OLIVE WHICHER

Goethean Science Foundation
Whitsuntide, 1952

INTRODUCTION

To the Second Edition

The Living Plant and *The Plant between Sun and Earth* were soon out of print. The coloured illustrations had been prepared for a German translation, but in 1960 a considerably longer work matured in the German language, needing more illustrations. This book, *Die Pflanze in Raum und Gegenraum*, came out in black and white, and the coloured illustrations were published in a folder, *Pflanze, Sonne, Erde* (*Plant, Sun and Earth*, 1963), with a short text in German and English (Verlag Freies Geistesleben, Stuttgart). *Die Pflanze in Raum und Gegenraum* has now reappeared, this time with 20 pages in colour, making possible an illustrated edition in English and one in French by Triades, Paris.

To this end, I have rewritten the English text, bringing it more in line with the German, though not so long. Readers who are familiar with the earlier English books will recognize large amounts of the already formulated text woven together with further text. The book is essentially still co-authored by George Adams and Olive Whicher; this I felt to be important and have striven towards it in every detail, following what I know to be George Adams' intentions. He died on March 30th 1963.

In the Notes I have included reference to the work of colleagues published since the first edition of this book, but it has not been possible for me to research in any degree into what may have appeared in the mathematical or otherwise scientific world at large.

Goethean Science Foundation
Hoathly Hill, West Hoathly
West Sussex.
Michaelmas, 1979

OLIVE WHICHER

COLOUR PLATES

COLOUR PLATES

III

IV

V

VI

VII

VIII

IX

XI

XII

XIII

XIV

XV

FOXGLOVE

BASELLA ALBA

XVI

XVII

XVIII

XIX

XX

Chapter I

THE LANGUAGE OF PLANTS

1 Plant Form

What is the magical secret which is spoken silently yet eloquently from the heart of every flower? Hidden in the undergrowth or flaunted high upon the hedgerow, a message is for ever being sounded if we could but hear it – a thousand modulations of one mighty theme.

Through the centuries Nature – above all, Plant Nature – has spoken to the heart of man. In glory of colour and individuality of form the plants speak. In the past, man listened to Nature as though in a dream; in recent centuries he has determined to become more consciously aware of her secrets. In so doing he has turned his attention especially to the inorganic world, and with the secrets wrested from this dark domain our modern life is largely fashioned. This dark domain is Pluto's world, and from it we hear indeed the rumblings of the deep.

Where Pluto reigns Persephone is chained – but only, so the legend tells, for one half of the year. Modern man in his quest of Nature's secrets has forgotten that when the plants speak to him in their living beauty of colour and of form it is of another world they tell than Pluto's realm. They tell of Persephone free, not bound.

In learning to understand *living* Nature as she speaks through the plant we must know the key to two worlds, the one in which Persephone is bound and also the other, where she is free. We must listen to the plant speaking from both these worlds, for the plant is indeed a synthesis of both, a creation like the rainbow poised between darkness and light. To understand this language consciously along the modern road of pure thought and scientific reasoning is not only possible but essential; it is vital for the future well-being of Nature herself and of mankind.

The plants have many qualities whereby we learn to know them. In colour, scent and substantiality they present themselves to us, each in a different way. Above all it is by virtue of its form that the plant reveals its being,

1   Hydrangea

2   Rowan

Enfolding Gesture at the Growing Point

3   Wayfaring Tree

4   Hazel (Plate I)

Younger and Older Leaves

5   Rhododendron (Plate II)

its specific character coming to expression most of all in the form of the flower.

What is a plant form? Often, perhaps in the very plants which give us most joy, the form is the most changeable, transient aspect of their being. As time passes the same plant may assume many forms, always recognizable yet ever changing. And the time may come when, as we look at the bare earth, we see no form at all. Yet the plant is still there; it is a seed, lost among the grains of soil, ready to spring forth true to type when the Sun is higher in the sky.

The plant lives between the darkness of earth and winter and the bright radiance of summer skies; it is ever there, never for a moment losing contact with its own individuality, yet expressing this so very differently at different times. To the winter, broadly speaking, belong the seed and the root, the unmanifest form of the plant, contracted and withdrawn into the darkness. In summer the plant-forms expand, leaf upon leaf and spreading branch in the light and air. Yet in every moment of this expanding, contraction still plays its part; at the node there is a bud in the axil of every outspread leaf. The unmanifest form awaiting development is always there beside the unfolded form.

Goethe saw the life of the higher plant as a threefold rhythm of expansion and contraction.[1] First, the expansion from the seed into leaf and leaf-bearing shoot; then the contraction into the calyx or involucre. Second, the expansion into the coloured petals, with the contraction into pistil and stamens. Third, the expansion of the fertilized ovary into the fruit, and the supreme contraction into the seed.

It is to this rhythmic life between contraction and expansion, darkness and light, that we must look, with Goethe, if we would further our understanding of the plant in a way which is true to its living nature.

2   *Leaves at the Growing-Point*

It is a wonderful paradox of Nature that the upward-shooting plant brings forth materials and forms which both in use and in appearance are proverbial for their radial penetrating power, yet there is little of this quality in the way they first come into being. The upright stem does not *thrust* its way into space like an arrow or a spearhead. The upward-growing power of the shoot is indeed one of the mightiest phenomena we know, and the eventual outcome of it is a thing of strength in the realm of earthly pressures and tensions – formed into pillar and pile, spoke and ramrod for human use from ages past. Yet it was not with this earthly-radial quality that the growing shoot made its way up and outward.

Not only are the young growing tissues delicate and watery; the same is true of nearly all living, growing things, including the downward-tending root which has indeed a radial quality of form and growth. It is not merely the delicate material; it is the form, the gesture of the growing shoot to which we specially refer. Describing it exactly as we see it, the typical phenomenon at the growing-point is the very opposite of a spearhead. What we behold at the tip of the

6 *Blackberry*

7 *Woodruff*

More Gestures at the Growing Point

8 *Goosegrass or Cleavers*

9 *Figwort*

growing shoot is concave and not convex; it is a hollow space we nearly always see. The actual growing-point of the stem is deeply hidden amid the young enfolding leaves. Their gesture is as if to guard, there in the empty space between them, a hidden treasure with protecting hands. The youngest leaves reach upward, sometimes in pairs and very close together, sometimes in whorls forming a hollow cone, first deep and steep, thence gradually opening and flattening. Often each single leaf is concave on its inner side, making the hollow space more spherical and cup-like (see Figs. 1–9).

Leaf after leaf, whorl after whorl with further growth expands and comes away, opening more or less towards the horizontal; meanwhile within them, other, younger buds have grown to take their place. So long as the shoot is growing, the gently guarded hollow space is there. It is a characteristic gesture that delights us, in all varieties of eloquence and beauty, all through the spring and early summer. Passing the height of summer, when we see less and less of the upward and enfolding gesture of young leaves but nearly all have opened full and flat, we know that this year's outward growth is ending.

This concave quality of upward growth is an essential feature of the impression we receive from the green plants that bedeck the Earth around us. The plants live by the light, coming to Earth from the Sun, from cosmic spaces. Pictorially, it is as though each single shoot were reaching out to receive and hold its portion of the light. All this contributes to the peculiar feeling of freshness and buoyancy the plant-world gives us. Leaf and leaf-bearing branch, as they grow older, do indeed tend out towards a planar and even horizontally flattened form; yet for the most part they retain something of the up- and inward gesture. Where the leaf does not flatten to the likeness of a plane, it is indeed not always but in the majority of cases concave on its upper ventral surface.

How characteristic this impression is, we may also tell from what we feel when it is absent, or when the ageing portions of a plant have lost it. If the plant lacks the hollow gesture altogether – like many cactus-forms for instance – it looks ungainly, quaint and untypical amid the higher plants.

3   Leaf and Plane

With the unfolding of leaf and branch is associated another quality which we perceive and feel in the phenomenon of plant-life above the soil, though science hitherto has lacked the corresponding concept. The leaves, as we said, tend to unfold towards a plane. They are, in a sense, *planar organs*. It is not only the crude quantitative fact that they develop a far greater surface-area than thickness; in their whole quality, function and morphological gesture they reveal that the character of "plane" belongs to them, just as the character of "point" belongs to every earthly object by virtue of its mass and weight – namely its centre of gravity.

We know the quality of "plane" first and foremost in the horizontal surface of the Earth around us. (Ideally, it is the tangent plane to the Earth's sphere at the particular point where we are.) We experience it if ever we look out over a wide open plain or over the still surface of a lake. And now in countless instances the fully opened leaves of plants – often the branches too, which bear them – make manifest this horizontal plane, or rather many parallel planes, one above the other. We see it when the sunlight falls through the young leaves in the beechwoods in May and June. A myriad planes seem to hover in the sunlit air. The impression we thus receive from the outspreading leaves is one of buoyancy and lightness. They seem to be upborne. This sense of buoyancy contributes to the feeling of life and refreshment which makes us glad to bring not only flowers but green leaves and twigs into our dwellings.

When we are looking through clear water at aquatic plants, we see the shoots buoyed up by the surrounding element and can interpret the phenomenon by the well-known law of hydrostatics. Yet for our vision and spatial feeling there is undeniably a like quality of buoyancy – even a stronger, more active one – about the upward striving and outspreading gesture of the terrestrial plants, though there is here no dense material element which in the Archimedean sense would relieve them of their weight. These are again the pure phenomena for which we shall be seeking the ideal counterpart, the interpretation.

4   *"Convex and Concave" in the Forms of Growth*

The full significance of the concave gesture of leaves at the growing-point will dawn upon us if we compare the higher plants with other forms of life, both in the vegetable and in the animal kingdom. Life has its origin and home in the watery element and manifests itself in growth. The primary and simplest form is a sphere, living in a watery medium, filled with watery or semi-fluid living substance and differentiated from its surroundings by some form of skin or surface-layer. The living sphere grows, drinking in water and other substances at the expense of its surroundings. This is the primary phenomenon of convex, outward growth. We find it in the microscopic, cellular organization of all living and growing things, including of course the higher plant – root, shoot and leaf without exception. But the cellular growth is here subservient to macroscopic forms of life, visible to the naked eye and more significantly diverse.

Every material body must ipso facto be predominantly convex towards the surrounding world, for it must be contained within a finite radius and volume. A living body therefore takes its start from the simplest of convex forms – the more or less spherical shape of the seed or germ-cell – and in its outline as a whole, whatever hollowings and folds, ramifications, incisions and excrescences it may involve, it must have something of the self-containedness of a finite convex form. So has the tree for example, which when we see it in winter-time shows very beautifully its convex outline, though this is formed by myriads of twigs and branches through which we see the sky (Fig. 13). In effect, convex though it must be in this sense, being a thing in the material world, the higher living organism reveals the interplay of an opposite, a concave principle of form. Yet the way in which it does so is profoundly different in the plant and in the animal kingdom.

In the animal, it is the well-known principle of gastrulation – *invagination*. In the lower animals where the simplest archetype of this process is revealed, the primary, spherical form of growth, the blastula, is hollowed from one end and folded in upon itself. In the resulting gastrula the original interior of the sphere – known as the segmentation-cavity or "blastocoele" – has now become the space confined and often more or less obliterated between the outer and inner folds, while a new hollow space, the "enteron", is formed within the latter. This kind of "invagination" – the turning-outside-in, the hollowing of originally convex forms – is repeated again and again in the course of animal development and embryology. To a great extent the complex and yet closely knit animal body, with its internal organs, its convolutions and membranes folded back upon themselves, is by such means produced.

The plant appears the very opposite of this. Its essence is to grow ever outward. Yet here too, in the shoot, a concave principle of growth reveals itself, as we have seen. But this concavity tends in the opposite direction. It is not like the dark and inward process of invagination. Out in the light and air, even by the upward and outward growth of the young leaves, a hollow space is formed at the tip of the growing shoot; deeply and closely enfolded as it so often is, yet it is only formed to be progressively *un*folded. In fact the plant-shoot lives and grows – if we may coin this expression – by a perpetual process rather of *e-vagination*. Such is its characteristic.

The higher plant does not shoot forth with a mere earthly-thriving life, sending forth spherical or elongated organs that would seem merely to thrust their way outward into space. (Such indeed is the form of growth of many of the lower plants – the fungi and to some extent the algae.) It reaches outward to a hollow space, which it then gives away as it unfolds. We only fail to recognize this because the hollow space seems empty; as it were, there *is* nothing to unfold from. When the animal infolds its gastrula – forming its *Urdarm* as Haeckel called it, its archetypal stomach or intestine – we know at once what this signifies, for into such an organ food, for example, will be ingested. But when the plant – which, as we know, *gives* more than it takes in the economy of life – pours out as it were the hollow space which it at first enfolded with such tender care, nothing is there to see but the surrounding light-filled air, and it requires an awakened insight to relate this very characteristic morphological gesture to the prime function of the green plants, which is to bestow life-giving oxygen and also nutritive substance both on themselves and on all other earthly creatures.

5   *Unfolding Growth and Plastic Outline*

The synthesis of concave and convex, upward- and downward-opening forms belongs to the peculiar magic of the higher plant, both herb and tree. The characteristic outline of a pine-tree – a Norway spruce for example – is of a cone widening downward from the apex, more or less dark and opaque against the sky (Fig. 13). Yet in the process and form of growth by which it comes into being we see the opposite: a hollow cone that opens *upward* – a gesture that begins already with the seedling and is repeated again and again in leaf and branch. This is a simple emblem of a more universal phenomenon. The opening from a deep hollow sphere or cone – characteristic of the youngest leaves and branches – expanding and flattening from thence towards the horizontal (Fig. 4), while simultaneously the apex of the stem and the plant's outline as a whole grows forth and outward to form a sphere- or cone-form, convex and as a rule widening downward from the apex: such is the growth of the plant in process and in outcome.

In the herbaceous plants with their more open growth the "outline as a whole" is often less in evidence, but the same principle will appear in the inflorescence, the development of which, as Goethe shows, involves an element of condensation – of contraction. We see the manifold, more or less closely packed inflorescences, conical as in the lupin, or spherical as in clover or hydrangea, or even more closely knit as in the composite family. Each single flower still has the cup- or tube-like form, only the inflorescence as a whole is convex. Yet the latter too was in its early stages deeply hidden – a mass of buds down in the hollow of protecting leaves. Thence it emerged, shooting up- and outward.

We may apply to this aspect of the growth of plants the ancient symbol of the two interlacing triangles – Solomon's seal – only we must change it from a dead static form into a functional, dynamic image. Multiply the upward-open triangle – we have the gesture of the growing leaves and branches. The single downward triangle typifies the form of the plant as a whole. As a two-dimensional image, this is already the type of many leaves with their triangular shape and spreading veins. Rotated about the vertical axis, we have the twofold cone-form, as in the pine (Fig. 13).

6   *Root and Shoot*

All that we commonly see of the plants around us are the leaf- and flower-bearing shoots; the roots are mostly hidden – do not become "phenomenon" until we dig them up. This too is significant, and it is not without meaning that Goethe the great phenomenalist, who derived so much of his perception from the sense of sight, says little of the root. The individual essence of the plant is manifested far more in the shoot; down in the dark community of Mother Earth there is not only the "mycorrhizal" but there are many other "associations"; this is far more of a *common* sphere of life. So too the botanist learns from the character of roots in general, but the roots tell him far less – comparatively speaking – about the individual character of one plant or another. We base our classification first and foremost on the flower, to a lesser extent on the leaf-bearing shoot, and least of all on the root.

Yet we must contemplate the plant as a whole – root and shoot – even to shed a

fuller light upon the form and gesture of the shoot itself. At the ground-level –
the region commonly known as the "hypocotyl",[2] where the root passes over
into the shoot – the plant is drawn together; here the cross-section is
comparatively small (Fig. 10) (Plate III). Thence it expands – downward and
upward, but in very different ways. Upward as we have seen, the plant achieves its
expanded space by a concave, enveloping, embracing form of growth. Even the
ultimately radial and convex forms – most prominent in the tree, which becomes
most earthly – come into being by means of this other mode of growth, the form
and gesture of which predominates in many a herbaceous plant throughout its
life.[3]

Downward on the other hand into the root, the growth is truly radial from the
beginning, and this is fitting, since the root enters far more deeply into the
earthly realm. The root-tip pushes its way through the soil; the living, growing
point (the so-called meristem) is covered by a root-cap, so that it comes to no
harm through this method of penetrating the earth. The primary earthy-
mechanical forces are radial – that is to say, working along a line from one point
to another, – wherefore our engineering structures so often take the form of a
network of girders, "struts and ties". Taproot reaching vertically downward, or
wide and ramifying system, the very form of the root suggests its adaptation to a
realm in which the material and earthly forces are most at home (Fig. 14; see also
Fig. 65).

We find a corresponding difference between stem and root in internal
structure. Thus at the hypocotyl there is a peculiar interchange of what is inner
and outer.[4] The concave and enveloping tendency of the shoot also finds
expression in the *cylindrical* character of the stem, which, if not actually hollow,
develops its essential organs, vascular bundles and cambium, in a cylinder
around the pith or woody trunk. It is from this outer enveloping portion of the
stem that the side-shoots spring. In the primary root on the other hand the
vascular system is internal, forming a central core, and the lateral roots spring
from this inner region, forcing their way through the outer layers (Fig. 10).

7   *Radial below, Enveloping above*

Sometimes a craftsman makes a wineglass form of cup or chalice rather like an
emblem of this twofold nature of the plant – perhaps suggested by it. The slender
stem must be supported on a wider base, and if this is chased and moulded on a
more radial pattern as it often is, we have the dual form in question – radial
below, enveloping above (Fig. 16).

But the plant-shoot is a plastic, ever-changing growth – at least until it comes
to flower. It forms the gesture of a chalice, as we have seen, only to open it out,
forming another cup within the first. So it repeats the process, rhythmically,
node by node (Figs. 11, 12). Moreover it can only do this inasmuch as the older
nodes are progressively thickened and strengthened, so as to bear the younger
and more delicate portion of the shoot ever upward. The lower portions of the

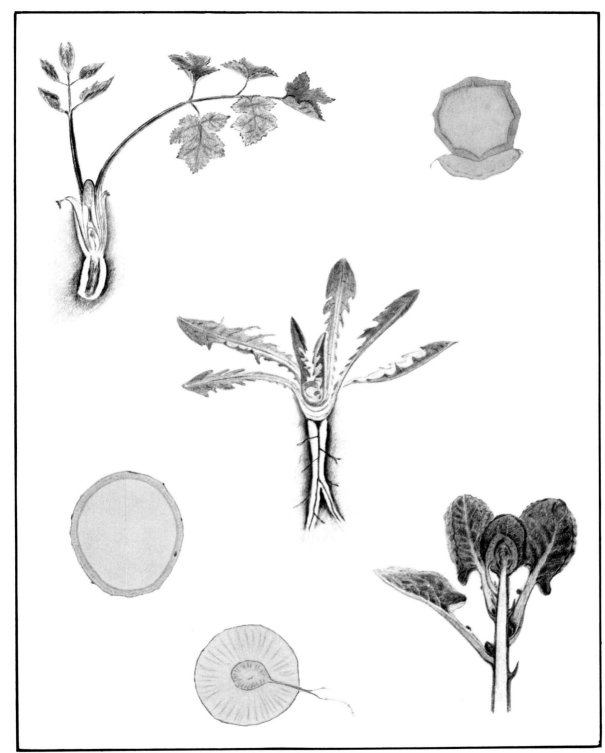

11   Silphium perfoliatum

12   Sheathing Petioles of Hogweed

13   Unfolding Growth and Plastic Outline

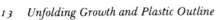

14   Beetroot

stem partly take on the supporting function of the root; they act as mediators between the actual root and the younger more living organs. To some extent there pertains to every node a modicum of the same function which belongs to the hypocotyl. How easily, in many plants, the node reveals this potentiality; transplanted under the right conditions, it will develop root as well as shoot.

Thus is the dual relationship rhythmicized and repeated. Botanists have compared the "morphological polarity" of the plant to the polarity of a magnet; the analogy is of course unsatisfying, for there is not enough qualitative difference between the north- and south-seeking poles. In the next chapters we shall see how a far more qualitative, morphologically true polarity may be discerned. For the moment, however, accepting the rough-and-ready picture, we may recall how a bar-magnet can be broken ever so many times along its length; each fragment will reveal the same polarity as the original whole. Moreover, given the opportunity, they will attract each other, end to end, tending to restore the whole both in form and function. Potentially it is as though the single magnet were already divided into ever so many partial magnets, all along its length. Something like this is the plant, but in a far more living way – with the caesuras marked not by arbitrary breaking, rather by rhythmic integration.

8   Cup within Cup

It follows therefore that the upward-opening and concave gesture by which a lower portion, rooted in the ground, carries the upper like a cup or chalice, pertains not only to the plant as a whole but is at least potentially repeated from node to node, cup within cup. Often, however, this relationship would only become visible if, so to speak, in the mind's eye we could add a time-dimension – if we could quickly and imaginatively follow the life of the plant backward in time and see the opened leaves at the lower nodes returning to their erstwhile gesture.

If this be a true reading, we shall expect to find that in some plants at least it becomes phenomenon. Indeed the cup-within-cup, hollow-within-hollow quality is very often in evidence. On a rather primitive level it appears in the peculiar segmentation of the equisetum-shoot, each segment springing from within the serrated leaf-sheath of the last. We see it in the habit of the grasses, where the lower portion of each leaf forms a sheath about the stem – or about the sheaths of younger leaves – reaching far up to where the ligule is. We see it too in the frequent tendency of the base of the petiole or leaf-stalk to envelop the parent stem, so that each internode of the latter grows out of an embracing hollow. Most eloquent, where a pair of opposite leaves arises at each node, are the connate forms such as we find in certain leaves of honeysuckle, or as a deep vessel in the teasel. (Figs. 81, 87, 88).

It is an archetypal form – this tendency of each successive shoot or internode to spring from within a hollow sphere, borne by the last. Only the extent to which it is revealed, varies from plant to plant; sometimes there is little outward

15   White Campion

16   Snowdrop

17   White Campion Seed Vessels

18   Brazil Nut

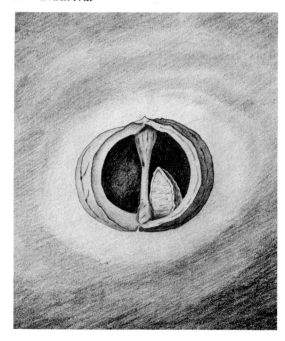

sign. In the Woodruff (Fig. 7) this typical growth-process is beautifully revealed.

Very typical in this connection are the plants – like Daisy, Dandelion, even Plantain – where a tall flower-bearing stem springs from a rosette of leaves at the hypocotyl. Though older leaves may be lying flat against the ground, such a rosette, taken as a whole, nearly always has a concave and embracing gesture, as if towards an ideal sphere into the focus of which the flower grows. What this type of plant shows as it grows from the earth, revealing a hollow cone or bowl or saucer of green leaves, out of which spring the flowering stems, others transform into a repeated rhythmic process. It is then a rhythmic process, which the flower at the top of the stem brings to a close. The flower cup is final, for then a quite new process sets in, leading to fruit and seed (Figs. 15, 17).

9   *The Flower's Chalice*

At the tip of a vegetative shoot the nodes and internodes are crowded together; often the leaves of many nodes combine to enfold the hollow space above the growing-point. Internodes lengthen out quickly as the plant shoots upward, but at the tip a younger sequence of unfolding leaves maintains the form (Plate XIII).

When in its further development the plant comes to flower, the gesture of a hollow and enfolded space is all the more enhanced. The changing of foliage-into flower-leaves goes hand in hand with a relative cessation of upward growth; many potential internodes are whorled together and cease to lengthen out. The flower-bud enwraps a space more tightly closed, and when we see it open to a flower it is as though the space were now poised in silence. What the growing tip of the vegetative shoot suggested in an ever-changing form – in an enfolded space, ever unfolded and renewed from within – this is now brought to rest in the flower-chalice, maintained as long as the blossom lasts. And in this "chalice", something hitherto unmanifest about the plant is now revealed, both in the form and pattern of the flower, by which the classifying botanist mostly discerns the individuality and type of plant, and above all in colour, texture, fragrance. One is inclined to say: if the hollow space tended by the young unfolding leaves, going before the apex of the stem as it grew upward, was not just emptiness but had a deeper meaning, its presence indicating a real sphere of forces scientifically still to be defined, then in the flower something more of this ideal space has been made visible. What hitherto induced the upward and unfolding growth, yet in its quality remaining latent, has now revealed its essence in another way. The material, sense-perceptible part of the plant has united with it more deeply than hitherto (Plate XV).

The flower, too, often unfolds its petals to a plane or even turns them back. Again, the hollow may be deeply exaggerated to a bell or tube, or sundry forms of nectary and spur. Yet in the main the cup-like form – in every degree of openness – is the most frequent and to imagination the most typical. If we have watched the enfolding and unfolding gesture of the green leaf-bearing shoot, our feeling of the flower-form's significance is all the more enhanced.

In our description we have now followed two of the three Goethean cycles of expansion. *One*: the unfolding of successive leaves and leaf-bearing side-shoots from the hollow space above the stem-tip – tiny at first, already adumbrated in the relation of cotyledon and plumule, and then expanding, undergoing variation, yet in essential quality remaining constant. *Two*: the unfolding of the flower, in quality quite different and preceded by a potent phase of contraction in the flower-bud and calyx. When at long last the flower-bud unfolds, the new expansion in its purely spatial, quantitative aspect is as a rule less lavish and less free; it is an opening rather of glory, a showing-forth of the plant's individual essence and essential beauty. Such is the second expansion – no mere variation but a true "metamorphosis", – in quality quite different from the first.

When we now come to Goethe's third and last expansion – the swelling of the fruit – we find an even greater change. Now for the first time the higher plant, in the shoot, attains a predominantly *convex* form of growth (Fig. 15). It is as though the hollow sphere, hitherto so immaterial, its presence only indicated by the enfolding leaves or by the cup-like gesture of the flower, were now for the first time to be filled with substance. For at this third expansion it is no empty, airy space that opens out, nor does the plant extend its body merely one- or two-dimensionally as in the slender forms of stem and leaf, but now the sap and growing tissues fill the whole volume of the fruit. We have the typical sphere-form of the apple or its equivalent in many variations. Heavily laden, the shoots that hitherto reached upward, bearing the hollow space at the growing-point or the fragile flower, are now weighed down with their fruity burden. Or the fruits harden into capsules, typical three-dimensional forms reminiscent rather of the mineral kingdom, or like strong ornamental caskets made by human hand.

This is again the enriching paradox of the life of plants through the summer season. From delicate and unsubstantial forms, now with amazing quickness is produced the earthly store of fruit and grain, to be weighed and garnered.

11   The Spatial Matrix of Plant Life

There is another most elementary morphological aspect of plant life. Goethe saw it as the interplay of a vertical tendency with a spiral tendency in vegetative growth; it is in the realm also of what Ernst Haeckel once called "Pro-morphologie".[5] Every tangible entity in Nature is in the three-dimensional world; if then a living body has any regularity of pattern it will make manifest not only its own particular nature but also some aspect of the structure of space as such. If we now ask, to what aspect of spatial structure does the higher plant mainly belong, we are led to Goethe's vertical and peripheral relation.

The crystal-mineral world reveals two great types of spatial form. One is the triaxial, that is to say, typically three-dimensional. Three axes meet in any given centre. They may be at right angles and of equal length and function, as in the "cubical" system; crystals of this kind will be most simply related to the prevailing laws of universal space and movement. Or the three axes may be of diverse length and even with their angles diversely inclined; other crystal systems are the outcome. But side by side with this there is a different type – the hexagonal and trigonal. Here there is *one* main axis, perpendicular to which there is a plane – or sequence of parallel planes – containing not two but three further axes, at equal angles of 60° to one-another. Rock-crystal (quartz) and tourmaline are typical examples.

The type of space here shown, with its single axis and series of planes at right angles thereto, is archetypally related to the rotational and spiral forms of movement (rotation about the single axis, which, if translation along the axis is added, will produce a screw-spiral movement). It is symptomatic that the crystal of silica – one of the most universal of earthly substances – does actually beget this form of movement, for when polarized light is passed through it in the axial direction, the plane of polarization is spirally rotated. Moreover, as among climbing plants there are those that form left-handed spirals and others right-handed, so there are left- and right-handed quartz crystals.

It is surely significant that the plant belongs to this axial, rotational form of space. Think for a moment of the most undifferentiated spatial form – the sphere. Though it is deeply related to the three-dimensional structure of space, no particular axial directions are singled out; it does not matter in what direction we begin if, say, we wish to circumscribe a cube about it. But as soon as we bring the sphere into *rotation*, if only for a single moment, we must rotate it about a *particular* axis; immediately, as on a geographical globe, this axis and the planes of latitude at right angles to it emerge, governing the whole structure.

Our Universe abounds in rotational forms of movement; for the Earth these find expression in the daily, yearly and other periodic rhythms and in the apparent circling movements of the heavenly bodies around us. The plant, of all earthly creations the one that most expands towards the Universe and takes its life and time from thence, belongs precisely to this form of space. The significance of this is all the more enhanced when with the help of modern Geometry our concept of the underlying form is deepened.

The normal plant is rooted in the earth; its only movements are the slow and silent ones of growth. Yet through the days and nights and seasons it is ever surrounded by the circling movements of the moon and planets, of the sun and stars. It rises towards the zenith and unfolds its leaves to the horizon, mostly in regular and spiral sequence. It is at rest amid an ocean of circling movements; only the mathematical regularity of its pattern and form of growth is of the circling and spiral type. Or are the delicate nutational movements of the growing-point also significant in this connection? Every now and then, in abnormal forms of growth, it is as though the "spiritual staff" were to lose its silent power; growth is caught up as if into a vortex. Such for example is the valerian plant, of which a picture is included in some editions of Goethe's work (Fig. 76, paragraph 44).

Let us now look once more at the three stages of growth and development. In the higher as against the lower plant, it is as though there were a focus of life and growth which is not immediately claimed by the material, watery-earthly body. The plant tends upward to this focus, enveloping it with its green leaves, which unfold and come away from it in turn. The flower then envelopes it more closely, no longer growing outward and away but pausing, and at the same time seeming to come into a nearer relation with whatsoever has been hidden here – making it manifest in colour, form and beauty. Yet still the hollow space remains; the inner focus has not yet been claimed. Only when the flowering process and with it the "fertilization" is complete, does the fertilized ovary grow right up into the living focus, or draw the virtue of it down into its substance. Now comes the third expansion; this at long last is material, three-dimensional, filled from within – growth at this stage is convex.

The plant has waited till this third stage to claim for its material life and body what it has hitherto left free and open. And when we contemplate the whole sequence of these phenomena, we see that to this waiting, this refraining, it owes its light and airy beauty. When at long last it comes to fruit and seed, uniting its earthly substance with the ideal sphere which until now it left untouched, the outcome is a greater fragrance and individual variety of substance than is afforded by the lower and more rapidly fulfilled forms of life.

In the flower, the ovary, containing the ovules, is an enclosing, green cavity; above or around it, the petals open, forming the beautiful coloured hollow, with the stamens within it, their anthers bearing the pollen. It is usual in the higher plant for the pollen to ripen in the air and sunshine. The pollen-grain is at the very summit of the plant's achievement as it strives upward to the light. The plant has raised a tiny particle of its living matter into the realm of the corolla, destined to become the bearer of the most vital forces. Fertilization is effected in many ways, but in principle it is always the same: it is the union of the essential content of the pollen-grain with the ovule. In the moment of union of these two realms, the rhythmic process of unfoldment will no longer take place. For a time, in the seed, the life lies secretly, silent and inert; with the spring sunshine it will stir again. Persephone will arise. In the moist Earth, warmed and irradiated by the light of the Sun, the seed will germinate, root and shoot will grow forth again. The plant will speak once more.

Chapter II

SCIENCE OF THE FUTURE

13   Goethean Science and Modern Geometry

Our task will be to seek for *Ideas*, which correspond just as exactly to the phenomena of plant forms and growth as do the familiar laws of physics to the phenomena of inorganic nature. To this end we shall bring to bear on the morphology of plant life a realm of achievement of the spirit of modern thought which has hitherto run alongside natural science, without having had any great influence on it. This is the modern Synthetic or Projective Geometry, the development of which took place in the beginning of the nineteenth century. Today, many mathematicians consider this beautiful realm of geometry to be classical and already superseded; in fact, it is but the seed of quite new directions in mathematics, containing as it does potent possibilities in relation to living nature and the science of life in all aspects.

The new geometry has a close relationship to the quality of thinking of the great artist-scientist Goethe.[6] It cultivates a qualitative, picture-forming aspect of mathematics, which approaches closely the Goethean experience of nature. For Goethe seeks the explanation of something living, not merely in the logically thought-out relationship of cause and effect, but through what he calls *anschauende Urteilskraft*. The term which comes near the rendering of this idea into English is *perceptive judgement*, meaning: a perceiving of the truth within the whole, while observing, so as to reach the archetypal picture or *Idea*, to which the phenomenon relates.

It is to such a method that Projective Geometry is akin. Not only in the method, but in the content of the new geometry, we find significant possibilities of bridging the gulf between Goethe's qualitative and completely phenomenological way of approach to Nature and the natural science of our time, permeated as it is with so much mathematical thought. Modern Geometry provides a way of experiencing space and spatial forms, so that the importance Goethe attaches to *polarity* in nature – light and darkness in the *Theory of Colours*, expansion and contraction in the *Metamorphosis of Plants* – may be approached through the transparency and exactitude of mathematical thought. Much that the biologist, who is imbued with the ideal of Goethe's theory of knowledge strives for, can be realized through the clarity and precision of studies in morphology based on modern projective geometry.

Unhappily, the one-sided insistence on the use of analytical mathematics as a tool for biologists has had a profoundly formative effect on the biological sciences and on the minds even of younger scientists, some of whom, however, know instinctively that a purely materialistic approach to the secrets of life yields no real progress, and are at their wits' end to find a way out of the impasse. The prevailing idea is that the inmost structure of matter must somehow contain the key to the phenomena – or else, if it does not, the key is not to be found. The biologist, beholding the wonderful regularities of pattern in the forms of life, tends to assume that if a rational explanation is to be found it must be via the atomic, ultra-microscopic realm with which the physicist and chemist are dealing. He is thus driven into a realm which in the quality of its forms is only indirectly and remotely akin to the phenomena of life directly visible to his senses. A striving to perceive the phenomena of life through the whole, rather than through the part, receives no help from the ancient, Euclidean, finite geometry inherited from the past. This is why there is a tendency in biology to borrow basic ideas from physics, for though in general the old conception of space is adequate for the understanding of inorganic nature, it is so only to a certain limit today. It may be said that the atomic physicist allows himself greater ideal freedom than the biologist. This dependence upon physics has undoubtedly been a hindrance to the proper development of biology. It has even been said that while biology in its effort to be an exact science has taken the basis of its ideas from physics, in future the laws of physics would reveal themselves to be special cases of the more universal biological laws awaiting discovery in the future.[7]

The general lack of interest and understanding of the potential which is inherent in the thought-forms of the new geometry results in a lack of the right kind of mathematical tool to complement the old forms and lead to a truer understanding of organic processes. Yet through the discovery in modern geometry of the Principle of Duality – better described as the *Principle of Polarity* – which bears on all ideas concerning space and spatial formations, the relationship of "Centre and Periphery" can be qualitatively experienced in quite a different way than before, and the needs of biology are met with an exact, scientific mode of thought.

Take for example the puzzling question of the morphogenetic processes in the early stages of embryonic development, which seem to be determined not from the inside of the material organism, but from the surface, taking place from the periphery inward. Gurwitch seeks for the Idea of peripheral, formative fields, partly enveloping the organism and working on its surface.[8] "Study" he says "the processes of growth and embryonic development in their early stages – invaginations, folding and the like – and we shall find the typical formations determined without exception by the contours of the outer surface, not by the internal structure." He and others have tried to deduce the specific character of the "morphogenetic field" purely from the phenomena of growth and form, leaving open the question of its source. Stress is laid on the surface-character of many typical phenomena of development, suggesting that the formative factor does not proceed, like the kind of causation with which we are most familiar in the inorganic world, from within outward, but the other way round. The elaboration of the "field" concept in physics encouraged scientists to look in this direction and attempts are naturally made to ascribe the field to the physico-chemical factors observed in the cells and tissues. The interrelation of cause and effect is, however, by no means clear, and the whole realm is wide open to research.

For such problems and phenomena of the living world, the Principle of Polarity in modern geometry provides a new and essential key; it awakens in the mind quite other powers of imaginative insight. Expressed briefly, the polarities of pure geometry suggest: *Wherever the point, there too the plane* – the plane which in contrast to the teeming world of atoms is related to the vast expanse of the spatial cosmos. Over against the point-centred elements hitherto known to physics, we are led to consider the thought of elements or processes planar in character, which by their very nature are akin to the celestial periphery, even as every centre of gravity is akin to the centre of the Earth. A schooling in this geometrical discipline gives to the mind capacities in thinking of a unique kind when applied to the observation of Nature; it offers the inclusion of the *plane* and of planes moving in from the periphery, as form- and space-creative entities, side by side with the *point* and its centric and contracted nature. The Goethean terms of "expansion and contraction" acquire a deeper meaning than the merely spatial one, for the Principle of Polarity leads beyond the conception implied by these terms in the old geometry. The law provides the *leitmotif* for the idea of metamorphosis. *Metamorphosis is possible in the changing interplay of polarities.*

14   Universal Forces – Rudolf Steiner's Indications

Goethe's achievements in biology, as well as in the theory of colours and with this leading into realms of physics, are being taken far more seriously in recent decades. Yet humanity today is in a very different situation, and is struggling both in natural science and in regard to the social questions arising from it, with problems hardly to be imagined in Goethe's lifetime. Until the end of the nineteenth century, the revelations of natural science were concerned with the material world. Deep spiritual powers lie at the basis of this realm of outer phenomena. "Space, Time and Matter" do not in fact form an absolute ground for science. Modern physics has come so far as to dissolve the seemingly stable aspect of matter, and to penetrate this, but in a one-sided and therefore certainly dangerous manner. *Space itself, in which matter has its existence, is in reality the result of the polar interplay of centric and peripheral components.* The recognition of this, which is reached with perfect clarity in the new geometry, has, however, not yet had the right and proper influence which it should have had upon natural science.

Man in his researches into nature's laws asks earnest questions; the universe answers him according to the nature of his questions. The one-sidedly centric forms of thought which have dominated physics and chemistry since the seventeenth century have opened up more and more the idea of the centric (atomistic) component of natural existence. Twentieth-century physics has long recognized that this component is of the nature of pure thought, and is not to be grasped with ideas derived from sense perception, as though the atom were a minute particle of matter. But there is as yet no recognition of the fact that natural existence also contains the polar aspect, even in matter and the substance-creating processes. *To the centric, pointwise component belongs the other pole – the peripheral component.* The ideas derived from modern geometry make it abundantly logical at least to widen the form of the question we put to nature. The peripheral component has to do with those cosmic powers, which in past times were experienced as the cosmic, ethereal and therewith the life-giving forces, the echoes of which are still to be found, particularly in Eastern traditional wisdom, though not accepted by modern science. It becomes more and more essential for the healthy development of mankind's culture that the knowledge of such cosmic forces be renewed, but in a modern scientific way. Rudolf Steiner[9] showed how to this end the necessary capacities and powers of knowledge can and must be developed on the sound basis of natural-scientific thinking and research. Especially towards the end of his lifetime, in response to questions from trained scientists, he gave indications concerning the ways in which a dominantly materialistic science must be transformed. Thoughts inherently related and applicable to the living processes have to be developed and applied in the observation of life, just as this has been done for the inorganic sciences. The laws of the inorganic world alone are just not sufficient to explain the secrets of the organic, and we must pass on from a stage in science in which far-reaching results have been achieved in mechanics and physics (and also in the mechanical aspects of medicine, in surgery) to a scientific era in which new thought-forms may be found, with which to approach more truly the physics and chemistry of the world of life.

A paragraph formulated and written down by Rudolf Steiner in March 1924,[10] indicates clearly the path which must be trodden by science:

"When we look out on lifeless Nature, we find a world full of inner relationships and find in them the content of the 'Laws of Nature'. We find, moreover, that by virtue of these laws lifeless Nature forms a connected whole with the entire Earth. We may now pass from this earthly connection, which rules in all lifeless things, to contemplate the living world of plants. We see how the universe beyond the Earth sends in from distances of space the forces which draw the Living forth out of the womb of the Lifeless. In all living things we are made aware of an element of being which, freeing itself from the mere earthly connection, makes manifest the forces that work down on to the Earth from realms of cosmic space. As in the eye we become aware of the luminous object which confronts it, so in the tiniest plant we are made aware of the nature of the Light from beyond the Earth. Through this ascent in contemplation, we can perceive the difference of the earthly and physical which holds sway in the lifeless world, from the extra-earthly and ethereal which abounds in all living things.''

On the one hand the study of lifeless nature has resulted in the formulation of the "Laws" according to which things relate to one another on and around the Earth; the stone rolls down the mountainside, the pendulum swings to and fro, the apple falls to the ground.

But how, indeed, does the apple get to the top of the tree in the first place? Is not the fact that it falls a secondary, rather than a primal phenomenon in the household of Nature?

In living processes, phenomena are to be seen which take place in an altogether different and usually quite opposite way from those which are familiar in the inorganic world. Substances move about in living tissues, rise upward as well as flowing downward; they may even interpenetrate, without losing their identity, which is impossible in the world of solid bodies. What is actually happening, when the rootlet pierces the integument of the seed, reaches down into the soil and draws substances upward from the dark Earth? How do the changes take place in the living tissues, allowing the young shoot to grow upward, instead of downward? Is the process called photo-synthesis fully understood in all detail?

What is it we see when, in Rudolf Steiner's words we "contemplate the world of living plants"? For the greater majority of trained scientists, the passage quoted above will be nonsensical, even embarrassing; and yet science today is asking questions, which, not so very long ago, would have been considered to be impossible. Here it is said: "In all living things we are made aware of an element of being which, freeing itself from the mere earthly connection, makes manifest the forces that work down on to the Earth from realms of cosmic space." What can this mean? How are forces that work down on to the Earth from realms of cosmic space *made manifest*? And what is the nature of such forces? Such questions are crucial today; they are being asked, and answers are being bandied about on all hands. But still more crucial for the life and well-being of man and the planet Earth is that the questions are rightly asked, that man researching into nature's laws asks the right questions and is ready to receive the right answers.

Rudolf Steiner took up and developed further the scientific method of Goethe. As a young man he edited and annotated Goethe's scientific works[1] and in the essay entitled *The Nature and Significance of Goethe's Writings on Organic Morphology*, he refers to and explains Goethe's description of the capacity by means of which organic nature may be comprehended, which he called *anschauende Urteilskraft* – the perceptive power of thinking (cf. p. 35). Goethe established in his morphological works the theoretical basis and the method of the study of organic forms and processes. The qualities perceived by the senses in a living form are the result of something which is not perceptible to the senses. What is perceived by the senses is not by itself sufficient to explain the total phenomenon. To do this it is necessary conceptually to grasp the whole – the Idea of the totality – as well as that which appears to the senses in space and time.

Such a morphological study requires intensive training in observation and thinking. To an awakened mind, the very forms themselves, their changing shapes and gestures "make manifest" their source and origin. It is the kind of knowledge, reaching to the impulsating, all-sustaining principle of life, which Spinoza called *scientia intuitiva*.[11]

In the areas of anthroposophical scientific work, based on the indications given by Rudolf Steiner particularly towards the end of his lifetime, research has been going on for some decades into the organic realm, attempting to demonstrate the etheric or ethereal forces. In comparison with the "centric" forces of classical physics, they have been called by Rudolf Steiner the "universal" or "cosmic" forces.[12]

The vast field of phenomena revealed by modern science, for instance in anatomy, zoology, botany, embryology, in earth-sciences and so on, can be approached with renewed insight. Experimentally, new techniques are being evolved and developed further in order to demonstrate the action of ethereal, cosmic forces in relation, for instance, to the effect of moon and other cosmic rhythms, or of high potencies, on living organisms. Research is undertaken into processes which are known, but are without scientific explanation, because the idea of peripheral or universal forces is lacking. Such experiments are being carried out particularly as the result of the practical needs of medicine, pharmacy and agriculture, where the healing aspect of Rudolf Steiner's indications is being more and more recognized today.

Although Rudolf Steiner gave many indications as to how to proceed in these various realms, he made it clear that the rediscovery of the ethereal realm of life must take place in a modern scientific way, and must not be confused by vague, traditional conceptions of the nature of these forces. It is necessary to redefine in scientific terms the realm and nature of the forces that shape and sustain living forms. He showed how it will be possible to form a clear conception of the ethereal entities and forces, which, he contended, would open up a field of true research, and, in a higher sense, of direct perception.[13]

It is in this context that the present work relates to Rudolf Steiner's scientific aims, bringing to the researches already in progress modern mathematical thought-forms and procedures. The specific direction taken, which develops the idea of a peripheral type of space and formative process on the basis of modern projective geometry (cosmic as opposed to earth-space), is due to concrete indications Rudolf Steiner gave, when he asked for the mathematical formulation of what he called "Negative or Counter-Space" (Gegenraum). As well as using the word "cosmic", he often referred to "etheric space" and to "sun-space" in the same context. He repeatedly pointed out the need to complement or transmute the conception of space, which is ideally formalized by the three cartesian axes, in order to overcome or supersede the one-sidedly centric and physically spatial thought-forms of natural science. Hence the formulation of a polar Euclidean type of space by George Adams (Kaufmann), which he published first in German under the title: *Von dem Aetherischen Raume* and then in English: *Physical and Ethereal Spaces*, both in 1933.[14] Later, the same concept was worked out by Professor Louis Locher-Ernst, called by him "Polar Euclidean Space" and published in 1957 in his book *Raum und Gegenraum*.[15]

Thus in recent decades the further steps have been taken in projective geometry, which set forth the mathematical statement of spaces absolutely complementary to the rigid and finite space of Euclid. Projective geometry had discovered that the ideal structure of three-dimensional space does not proceed one-sidedly from the point alone, but from two opposite entities – point and plane – which play a fully equivalent part in the fundamental structure. Already in the 1820's and 30's the "non-Euclidean" geometries had been discovered; moreover, spaces of more than three dimensions had been thought out and the classical *a priori* of a rigid and right-angled three-dimensional space as being the only possible way of thinking about space was shaken. A still further change was wrought by the increasing effort of nineteenth-century science to penetrate into the working of Nature's forces and to adapt the forms of thought to the phenomena discovered. No longer could researches concerning "real" objects and events in Space and Time remain within the framework of the classical concept of three-dimensional space, nor accord with Newton's idea of the uniform flow of Time. Thence came Einstein's Theory of Relativity and other subsequent developments. The idea of a conception of space *polar* to Euclidean space was therefore theoretically close at hand, but furthermore the way was also open to the clear mathematical concept of forces polar to the conventional, centrically conceived forces of physics.

Because of the one-sidedly mechanistic and materialistic direction of applied mathematics, which has served science well in the direction of atomism, it appears that no one, apart from the researchers following Rudolf Steiner's indications, has deemed it worth while to work out the formulation of polar Euclidean spaces, let alone to develop concepts concerning the forces which might apply in such a space. Yet this is precisely what is needed to give to the concept of etheric formative forces – processes at work in the building up of living tissues and forms – a clear mathematical basis. In other words, what may be revealed through observation by the process of "perceptive judgement", when an underlying truth or *Idea* has been perceived, can be described and explained in terms required by the modern scientist – provided, that is, that he has acquired the necessary mathematical skill and imagination.

In the present work the attempt is made to bring these new ideas to bear on the morphological phenomena of living nature, and in particular of the higher plant. Our underlying method is primarily morphological. In the observation of the living forms in all their detail and beauty, we shall be seeking to read the ideal – even mathematical – significance of the visible, macroscopic phenomena of plant life and function. Moreover, we shall approach the plant as a whole, – its changing forms, the way it creates its own spaces as it lives and grows through the seasons.

Our work relates to what can be seen with the naked eye; we shall leave aside the details of cell structure, the phenomenon of cell division, and so on, although, of course, we are conscious that this is a lack. It is, however, in the nature of the new method, to approach at first the whole and then its parts. According to the old spatial conceptions, it is natural to think first of the parts and then to put them together to make up the whole. One thought instinctively that as the whole is made up of smaller parts, it is therefore to be explained by a study of these parts. The cell-structure of the leaf revealed by the microscope, for instance, must surely contain the explanation as to how the whole leaf has come about. The structure of the protoplasm, seen by means of a stronger magnification as plastids, chloroplasts, chromoplasts, nucleus and so on, must, we tend to think, contain the explanation of the cell as a whole. Finally, one hopes to find explanations for these smaller particles by penetrating to the still smaller molecular structure with the electron microscope. This line of thought has indeed, particularly in recent times, led to important knowledge; but the deeper insight into the type of spatial concept afforded by modern geometry leaves the question open as to whether perhaps the explanation for microscopic phenomena may not just as well be sought in the phenomenon of the whole. Many phenomena point neither in the one nor in the other direction; to an ordinary spatial mode of observation it is indeed a puzzle that the same kind of cell-structure and process of cell-division should exist in the myriad forms of both animal and plant life. This fact alone might well be an indication of the insufficiency of this manner of explanation.

We shall be led by our study of plant life and also by the spirit of the new geometry, with its deeper conception of space, to experience spatial formations – even space itself – as evolving processes, rather than as being already given and ready made. This directs us towards a way of approach to knowledge concerning which one can become convinced that it will be just as characteristic of the science of the future as was the predominantly spatial mode of approach for science hitherto.

As well as seeing Nature as a series of objects and processes spread out in space and taking place in the indifferent passage of time, we shall begin to recognize Nature herself as a time-organism uniting earthly and cosmic polarities. Universal epochs of time are super-imposed one on another in Nature's laboratory. Just as in the human soul experiences of the present moment – joys and sorrows – live together with memories reaching back through years and decades, so it is in the space-time existence of Nature. Things which co-exist in space and in the immediate flow of present time – particularly when it is a question of different orders of size – can correspond to different cosmic times. It becomes a matter of recognizing the *signature of Time* at every stage. Then the explanation can no longer be a spatial one, not even one which, in the sense of Einstein, includes Time as a dimension within the spatial structure.

It is in this sense that a morphological study, which takes its start from the visible forms and growth processes of plants and leaves aside for the moment the histological and cytological phenomena, is justified. It must, however, be said that to investigate this realm also with the help of the ideas of Space and Counterspace, once the first steps have been laid in the study of the whole, will be an extremely important task. The question must be asked, how the unfolding development characteristic of a particular type of plant, say, a Rose, is related to the cell-structure and to the millions of cell-divisions which take place within its structure, how the one and the other aspect are mutually interdependent – how, perhaps, in the play and interplay of the Nature forces, the plant in its wholeness comes about not because of, but in spite of the cytological aspect.

Chapter III

THE POLAR FORMS OF SPACE

16 Perspective Transformations

The fundamental notions of projective geometry are not really difficult of attainment; they are less difficult and certainly far less abstruse than many of the mathematical ideas applied, for example, in the physics of our time, in which large numbers of people take an intelligent interest. These fundamental notions are, however, comparatively little known. To enter into them requires only a certain effort in active and imaginative thinking, such as does not always come easily to people today.

We shall introduce them here descriptively, as simply as possible, with the help of the illustrations, it being of paramount importance to activate one's picture-thinking in a mobile and qualitative way, thus evoking a healthy feeling of form. The relationships and metamorphoses are related to the living rhythms of our life, which penetrate not only our abstract thinking, but our whole being. The meaning and content of the geometrical truths will become apparent through description and illustration, but we shall leave aside proofs. Reference can be made by those requiring a systematic treatment of the geometry to the relevant publications to which reference is made in the Notes and References.[16]

Historically, projective geometry, the modern form of geometry, which transcends Euclid's ancient form, arose out of the *transformation* of geometrical figures. Their transformation to begin with, as the name implies, is by perspective, where a plane figure, for example, is "projected" from one plane into another. "Transformation" means "metamorphosis"; it is not surprising that this kind of geometry is akin to the Goethean morphology.

Figs. 19, 20 and 21 show, for example, the perspective transformation of the circle in the oblique plane (pictured as an ellipse) into the curves in the horizontal plane below. In such a way, the parabolic or hyperbolic form of light is thrown on the surface of the road by the cone of light from a lamp with a circular aperture. According to the angle of the cone of light, the circle will appear in a different form on the horizontal plane, appearing as one or other of the "conic sections" – we will call them "*circle-curves*": ellipse, parabola or

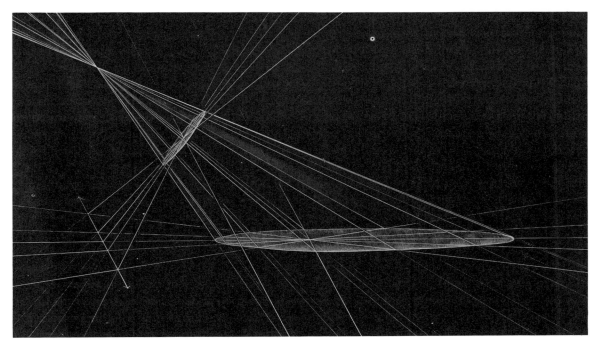

19　Perspective Transformation of Circle into Ellipse

20　Perspective Transformation of Circle into Parabola

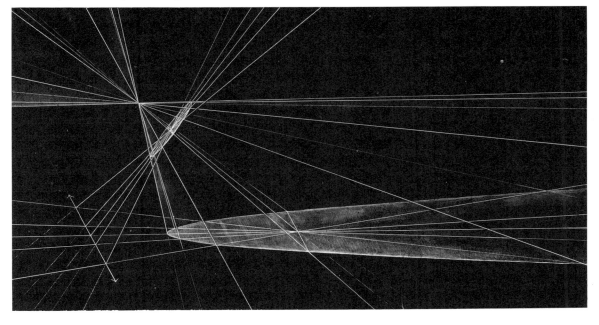

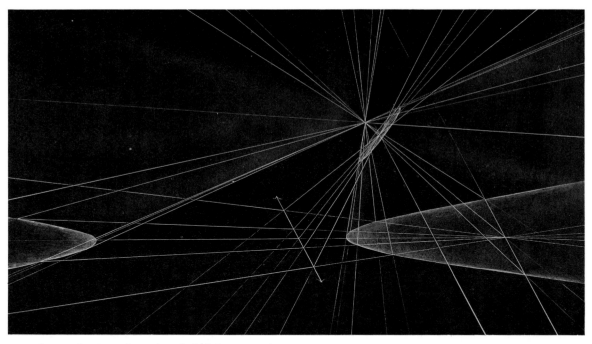

21   Perspective Transformation of Circle into Hyperbola

22   Family of Conics in Oblique Perspective

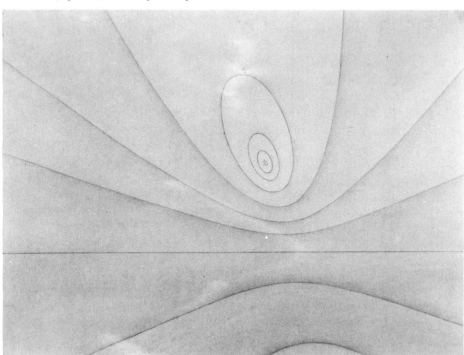

hyperbola.[17] The parabola appears when *one* of the rays (a so-called generator) of the light-cone is parallel to the horizontal plane. Set the projecting eye-point a little higher up and the parabola changes to an ellipse, a little lower down and it changes to a hyperbola, the ray which pointed to the infinitely distant point of the parabola having inclined to the other side of the oblique plane. Whatever properties remain unaltered by the transformation are held to belong to the form, as it were, upon a deeper level; they are more fundamental than those that suffer alteration, when the original aspect of the form is changed. A circle, for example, changed by perspective into an ellipse, no longer shows the constant radial distance from the centre. But there are other properties it shares in common not only with the ellipse but with the other, still more different curves into which it can be transformed, the parabola and the hyperbola. Fig. 22 shows, in an oblique perspective, the transformation of these curves one into the other in the sequence resulting from the movement of the raying point.

Fundamental is the theorem discovered by Blaise Pascal around 1640, when he was only sixteen years of age, about the inscribed hexagon. Join any six points of any circle-curve to form a hexagon of any shape, and the three points common to "opposite" pairs of sides will be in a line (called the Pascal Line).* It is a remarkable phenomenon, capable of being expressed in an infinitude of ways, in which – it should be noted – there is never a question of measure, neither linear nor angular (Figs. 23 and 24). The constant element is simply the *relative positions* of the points and lines and their qualitative relationships brought about by the curve. When the curve with inscribed hexagon is projected, as in Fig. 19, from one picture-plane into another, each line of the first plane together with the eye-point forms a plane, which meets the second picture-plane in yet another line. The linear quality of the relationships is preserved in projection, and though measure and form are radically changed, the *idea* of the inscribed hexagon always remains.[18]

Modern geometry leads to a conception closely akin to Goethe's concept of an ideal Type. Each of the fundamental forms it recognizes (curves, surfaces and so on) has many manifestations, outwardly often very different from one another. The ideal form as such cannot be identified with any one of its aspects; it hovers, as it were, over and among them all, recognizable to pure thought. It is made manifest in all of them, yet fully manifest only to a "metamorphic" thinking, which can pass freely and intelligently from the one to the other. It is in this sense like Goethe's "archetypal leaf".

17   The Infinitely Distant Elements

The young dawn of the new geometry was at the very beginning of modern time – in the fifteenth to seventeenth centuries. It occurred in the first place in the form of a practical theory of perspective, in connection with the work of artists and architects. From this beginning there gradually developed the new field of

* The word "line" in the geometrical part of this book always means "straight line". (See Note 17.)

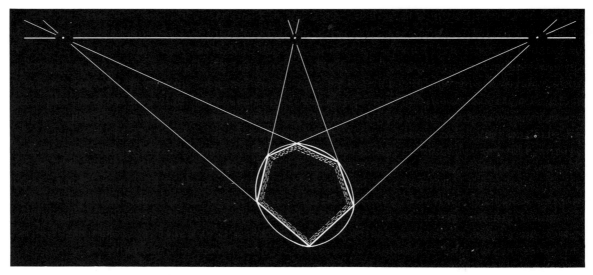

23   Theorem of Pascal on an Ellipse

24   Theorem of Pascal on a Hyperbola

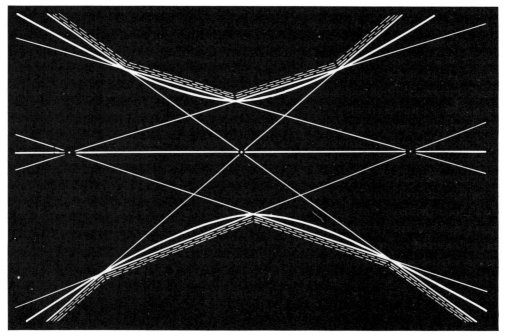

pure mathematical and geometrical thought. The actual sunrise was, however, at the beginning of the nineteenth century, still just in Goethe's time, and it then became more and more clear to the mathematicians that this discipline of projective geometry included and illuminated all the other known forms of geometry, revealing much which was foreign to the classical geometry of the old Greeks and Arabs.

Not only did geometrical thought come to terms with the infinitely distant elements – point, line and plane, but because of this very fact it became able for the first time to understand on a deeper level and to formulate the laws of polarity as expressed in point, line and plane and their mutual relationships.[19] This prepared the way for many possible conceptions of space to arise, beyond the classical Euclidean conception and including the idea of polar-Euclidean space or counterspace (Gegenraum), to which reference has already been made in the previous chapter.

For the geometry of Euclid, the point is minute, a centre of no dimensions, the line denotes direction and a definite distance in one dimension, while the plane is an extent of flat surface, an area stretching away two-dimensionally on all sides towards its boundary. In Euclidean geometry the primary element is the point; lines and planes and also volumes being built up from points by methods based on measure.

In modern geometry the three ideal entities interweave to create one another; they are in themselves formless, i.e. without measure. A drawing can only reveal them partially. Each entity may be thought of as existing in its own right, or as a "manifold" – an organism whose parts or members are the other two elements.

For example, the plane is woven of points and lines. In Fig. 25, the plane is shown consisting of a multitude of lines and the points in which they interweave. Obviously, only part of the plane can be depicted and only a few lines, the pattern of lines and points being extensible in all directions in the plane. The ideal plane is of infinite extent.

The point is a manifold of planes and lines. Figs. 26 and 27 show a point determined by interpenetrating planes and lines. The latter, though each of infinite extent, will all be *contained* in the one point. The ideal point is of infinite *content*.

The line is formed of points or of planes. In Fig. 28, both aspects of the line are illustrated; it may be formed by an infinitude of points *or* by an infinitude of planes.

In this way of thinking about and defining the three basic elements, *the plane is just as primary as the point*. Two planes are sufficient to create a line (as are two points); three planes, unless they are all in the same line, will always create a point. The lone point, shorn of its parts, which is the basis of Euclidean geometry, no longer has a place in the new geometry; it always appears embedded, as it were, in a matrix of lines and planes. Its members are the planes and lines which are in it, just as the plane is membered by the lines and points which lie in it.

It may be stated as follows:

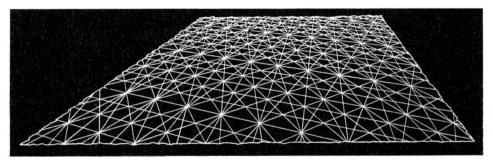

25 *Plane of Lines and Points*

26 *Point of Lines and Planes*

27 *Point of Lines*

28 *Line of Points and Line of Planes*

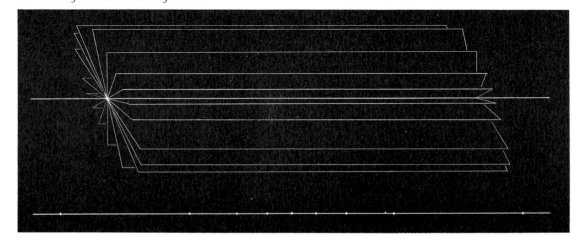

Line as a manifold of all the planes it contains.	Line as a manifold of all the points it contains.
Plane as a whole.	Point as a whole.
Plane as a manifold of all the lines it contains.	Point as a manifold of all the lines it contains.
Plane as a manifold of all the points it contains.	Point as a manifold of all the planes it contains.

As points are to planes, so are planes to points. The line is related equally to points and planes. Point, Line and Plane thus form a trinity, with point and plane representing the polar opposites, and line the intermediate, balancing factor.

The completeness of this formulation rests on the acceptance of the idea of the infinitely distant elements. For example, consider the statement concerning three planes: Any three planes not all in the same line will create a point (just as any three points not in a line will create a plane). Any two of the planes in Fig. 28 will create the line, and it is easy to picture a third plane striking through this line and thereby creating a point, as in Fig. 26. But now picture the third plane to be *parallel* to the common line of the first two; in this case the common point of the three planes will be at infinity. It is an ideal point.

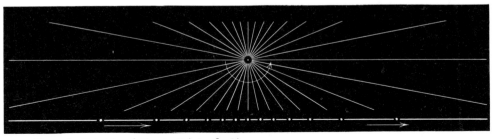

29   *Lines of a Point in relation to Points of a Line*

Fig. 29 is an aid towards grasping this conception: the two infinitely distant points in the two opposite directions of a line are identical. The lines of a fixed point ray out and relate in perspective each one with a point of a fixed line. Beginning with a point somewhere in the line and allowing the ray to turn in the direction of the arrow, the point on the fixed line will move away more and more quickly to the right. When the ray reaches the parallel position with respect to the fixed line, the point disappears momentarily *into* the infinite to the right, only to reappear in the next moment *out of* the infinite on the left, as the ray continues to turn.

To gain an impression of the nature of *the ideal line at infinity of a plane*, it is not difficult to picture this figure as a section of a spatial form. The fixed line becomes a resting plane, while the fixed point becomes an axis around which a whole plane turns; the planes have a common line, which moves parallel to itself out to the infinite (when the two planes are parallel) and back again.

Now give the revolving plane more freedom of movement, allowing it to pivot on a fixed point, inclining at all angles, without leaving this pivot-point. One can picture the common line of the two planes sweeping out to the infinite in a particular direction, only to merge into the *one infinitely distant straight line* of the fixed plane, each time the pivoting plane comes into the parallel position. It matters not in which direction of the plane the line is moving; it will merge in the last resort into the *same* line at infinity.

Such exercises of geometrical imagination help in the realization of what the mathematician means when he says: *The line-at-infinity of a plane is the locus of all the infinitely distant points of all the lines which lie in that plane.* Furthermore: *The plane-at-infinity of space is the locus of all the infinitely distant points of all the lines in space and of all the infinitely distant lines of all the planes in space.*

Thus, the ideal points, lines and plane at infinity are assumptions *in thought*, understandable and entirely reliable concepts, the validity of which can be demonstrated and proved.

18   Polarity: Point, Line, Plane – ("Principle of Duality")

It becomes obvious that in the membering of space there is a kind of symmetry in regard to point, line and plane and the idea begins to dawn that maybe there is justification in considering space itself and all spatial forms from both aspects – not only as formulations of points, but also of planes.

Thus it was that two French mathematicians of the beginning of the nineteenth century (Poncelet and Gergonne) discovered what they called the *Principle of Duality* which was then further developed by significant thinkers (Jacob Steiner, Chasles, v. Staudt, Plücker, Cayley, Grassmann and others). They called the space in which this balanced interplay of point, line and plane rules "*projective space*". It is the space which results when in inner imagination and pure thinking one does not consider the infinitely distant elements as being any different from all the others, they all have the same value. The forms in this space are changeable and never rigid. In this book we call this space the "*free, archetypal space*".[17] One might almost say, it exists as a potential in the pure light of thought. Forms become created within it through the projective process by means of the activity of thinking. It is like a matrix, a realm in which all possible forms exist, awaiting creation, and when once a form appears, it may take on all the possible aspects derived from the interplay of the archetypal entities, planes, lines and points.

The basic relationships between these elements may be described quite simply; we have seen that there is a balanced pairing of point and plane, with the line mediating between the two. They give rise to the axioms upon which the whole discipline of thought rests; these Axioms of Incidence may well be called the Axioms of *Community* of Point, Line and Plane in three-dimensional space:

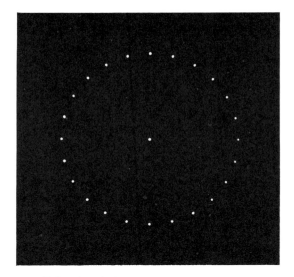

30   Pointwise Circle

31   Linewise Circle

32   Pointwise Parabola

33   Linewise Parabola

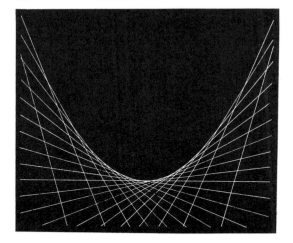

Any two planes have one and only one common line. This line contains all the points which the two planes have in common.	Any two points have one and only one common line. This line contains all the planes which the two points have in common.
A line and a plane always have a common point. If they have more than one, then the whole line with all its points lies in the plane.	A line and a point always have a common plane. If they have more than one, then the whole line with all its planes lies in the point.
Any three planes have a common point. If they have more than one point in common then all three lie in a line.	Any three points have a common plane. If they have more than one plane in common, then all three lie in a line.

Two lines either have *both* a point *and* a plane in common, or they have *neither* a point *nor* a plane in common. In the latter case we call them "skew".

For three-dimensional projective space, this may be briefly formulated: "As point is to plane, so is plane to point; as point is to line, so is plane to line." All the fundamental relationships of point, line and plane, evident to the imagination, can be stated in pairs, showing that point and plane play corresponding parts, with line related equally to either. Figures or propositions, mutually related in this way, are called the *dual* of one-another; the mental process of deriving the one from the other is called *dualizing*.

The principle might better be described as one of *Polarity*, even though the word has already been adopted in a more specialized sense (polarity with respect to a conic, quadric, etc.).

This archetypal reciprocity between the points and the planes of space manifests in all forms; for any assemblage of points in space there will be a reciprocal, planar form. Thus every form has an answering form. Moreover by "form" is meant not necessarily something rigid and permanent, but something changing and transforming – in other words, it can mean the metamorphosis of an ideal Type.

There is a very significant qualitative difference between a perspective transformation (they are called collineations) which shows gradual change from one extreme manifestation of the archetypal form to the other, and the kind of change of form which comes about when a form is turned into its polar opposite. This kind of transformation is *polar reciprocation* (correlation), in which a radical change is revealed between the original form and its transformed state, with no intermediary steps. In Figs. 19, 20 and 21, the sister forms of circle, ellipse, parabola and hyperbola are shown in three different pictures. Fig. 22 shows a whole family of sisters, revealing the perspective (projective) transformation between one form and another, such as would occur during the movement of the projecting point. Ideally there are of course an infinite number of curves between the point or focus on the horizontal plane and the line in which the two planes interpenetrate (directrix). In this homology type of transformation, all possible intermediary steps are ideally present. The transformation is very reminiscent of the variations – leaf-forms which can be seen up and down the stem of a plant, or between different varieties of the same plant.

19   Polarity called forth by the Sphere – Polar Transformations

The sphere – the purest and most ideal representation of the polar aspects of form as such – rules in space with a kind of active, balancing potentiality; it brings about the polar reciprocal type of transformation. This is understandable, when one considers that the sphere itself (and all its sister forms – or perspective transformations of itself) has two aspects, similar to the polar aspects of the curves in the plane (Figs. 30 to 33). The sphere, which is a spatial form, can be thought of as an assemblage of points, all equidistant from the central point; this is usually the only way we think of a sphere. In the new geometry it has, however, also to be understood as an assemblage of *planes*; ideally, to every point on the sphere's surface there is a tangent plane. The concept of a sphere is therefore only fully present, if we are able to think also of the assemblage of tangent planes, all of infinite extent, which envelop it on all sides, shaping its hollow form from the periphery inward. This is an unusual and perhaps difficult thought; it is, however, essential to the understanding of the idea of metamorphosis. The quality of transformation exercised by the sphere is the change from centric to peripheral types of form, where the relationship from one to the other may be quite unrecognizable outwardly. These forms are called "polar conjugate", an interesting use of terminology, meaning something quite different from a "family" of curves, such as concentric circles or their perspective transformations, as pictured in Fig. 22, and in Plate IX.

The understanding of the difference between these two basic projective processes can be an important and reliable guide in the study of organic morphology and in coming to terms with the idea of metamorphosis.

In three-dimensional space, the five regular polyhedra are beautiful examples of the polarity of point and plane. The cube in Fig. 34 is seen created in the interweaving of planes and lines, which bring about the points, and this is how we should think of the other, more complicated forms, with their many planes, lines and points. Rather than calling these forms the "Platonic solids", as is usually done, we prefer to call them the "Platonic *forms*", for we are not thinking only of a fixed form in its finite aspect, but of the form-types, considered also from the aspect of the plane. The forms consist of so and so many *plane* surfaces, which, interpenetrating one another, create the lines as edges and the points. The planes and lines are in reality of infinite extent, though the drawings only show the finite surfaces.

It is in fact the sphere which calls forth for a given form its polar reciprocal

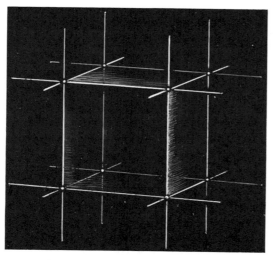

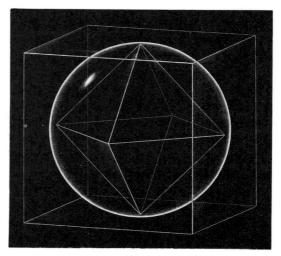

34   Cube formed of Planes, Lines and Points

*35   Polar Transformation (Metamorphosis):
Cube and Octahedron*

36   Cube contracting, Octahedron expanding

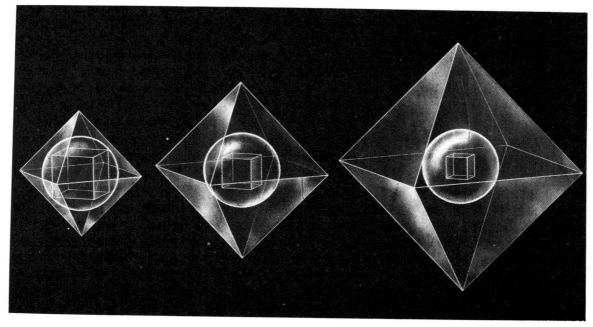

form. In Fig. 35, the sphere is set within the cube, so that the mid-point of each of the six faces of the cube touches the sphere's surface. These six points give rise to a second form, now within the sphere, – the octahedron. The two forms reveal a complete polarity in regard to point and plane; cube and octahedron show themselves to be polar conjugate forms. They arise out of one another by the mutual exchange of point and plane called forth by the sphere. In Fig. 36 (left), the roles are interchanged; the sphere is now inside the octahedron, and in the eight triangular planes of the octahedron there arise the eight corners of the cube, in which three lines and three planes meet.

Fig. 37 shows a similar pole and polar reciprocity with respect to the pentagon dodecahedron and the icosahedron, and Fig. 38 the self-polar forms of the tetrahedron.

These five forms are the only possible quite regular spatial forms, apart from the sphere itself, which, as we have seen, is also self-polar, like the tetrahedron, for we must think of the sphere from within *and* the sphere from without, – the pointwise and the planewise sphere. Put together, these *two* aspects reveal the whole concept *sphere*.

We have said that the sphere rules in space, balancing polarities. This is not only so when, as in the previous illustrations, the sphere *touches* the one form from within and the other from without. Let the cube, as shown in Fig. 36, diminish towards the mid-point, while the sphere remains unchanged. Through the relationship of cube-planes, -lines and -points to the sphere, there still arises an octahedron, for the sphere calls forth *to every plane in space its point or "pole" and to every point its "polar plane"* – the so-called law of pole and polar with respect to the sphere. To the extent that the cube leaves the surface of the sphere and diminishes towards the centre, the octahedron grows outward towards the periphery. Contraction and expansion keep pace with one another.

Figs. 39 and 40 help in the understanding of this process. In Fig. 39 (left), plane and point are co-incident, as in Fig. 35 in the case of the cube and octahedron. When, as in Fig. 39, the plane is raised above the sphere, its pole is determined by the tangent cone, which determines a circle of contact, which in turn determines a plane passing through the pole. The important Law of Pole and Polar with respect to the sphere is thus demonstrated; a point within the sphere is related to a particular plane outside it and a plane outside to a particular point within. Following the process to its ultimate extreme, it will be seen that when the plane outside the sphere reaches the plane at infinity, its pole will be at the centre of the sphere, forming a functional infinitude within. If on the other hand, the plane moves right in, penetrating the sphere, its pole will move outward towards the infinite, which it will reach, when its polar plane actually reaches the innermost centre (at which moment the tangent cone becomes a cylinder; in projective terms, it is a cone with its apex in the infinite).

Following the process taking place in Fig. 36, as the octahedron grows, its points and lines get further and further away, until they merge into the infinitely distant plane of space, and all its eight planes disappear into this plane at infinity. *There remain three points and three lines in the plane at infinity.* At the moment

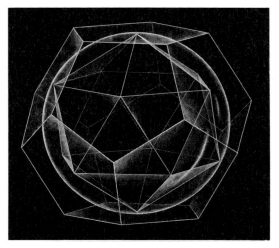

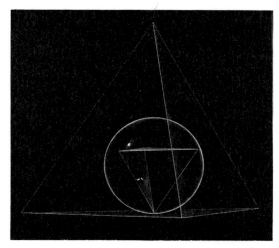

37   Icosahedron and Pentagon Dodecahedron   *38   Polar Tetrahedra*

40b   Polarity with respect to the Sphere

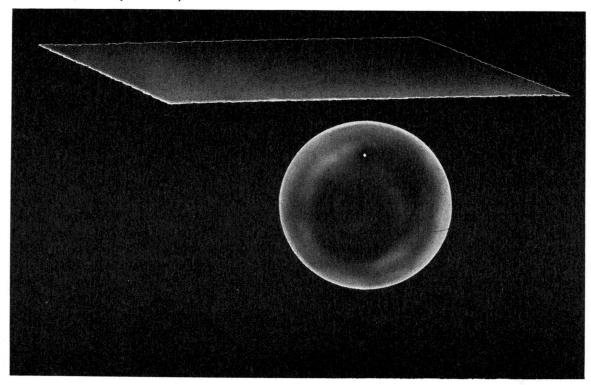

when this happens, the cube points disappear into the centre of the sphere and *there remain three planes and three lines in this central point.* Thus the ultimate polarity is between the plane at infinity of space containing three points and three lines (a triangle) and the central point, or *point at infinity within*, in the centre of the sphere, containing three planes and three lines (a trihedron).

It is important to try to visualize the polar conjugal form of the triangle – the trihedron – a form of infinite extent, opening out to the infinite in both directions, its three planes and lines all held in a single point. The law of "Pole and Polar with respect to the Sphere" is absolutely fundamental to all considerations of morphology and of space itself. Stated simply: the sphere (or any of its sister forms) calls forth, or accords, to each point in space a plane and to each plane a point.

Looking again at Fig. 39 the tangent cone is (partially) depicted, which determines the plane, which is polar to a point outside the sphere; this polar plane is determined by the circle of contact which the cone has with the sphere, and in this drawing, the plane, which is of course in reality of infinite extent, is only shown as an ellipse. Now move the point anywhere in the horizontal plane, and the cone will change its position, resulting in the movement of its circle of contact with the sphere and therefore of the polar plane. In the drawing, the point is marked upon which this plane would pivot, this point being the pole of the plane in which the apical point of the tangent cone is moving. So long as the moving point in the plane above remains in this plane, so long will the moving plane interpenetrating the sphere pivot upon and never leave the point within the sphere. It is from such considerations that the relationship of polar plane and pole is determined; it is an entirely mutual, reciprocal relationship, permeating the whole of space. Sphere, Polar Plane and Pole belong together as do parts of an organism.

Now take the thought one stage further. We have given the point in the horizontal plane complete freedom of movement *within the plane*; so its polar plane has complete freedom of movement *within the point or pole*. Thus, to whatever form the moving point might draw in the horizontal plane, there will be a polar reciprocal form in the point or pole. This is a very valuable thought form, or ideal picture, which turns out to be of unique importance in the study of plant morphology. It is very worth while gaining a facility in this type of mobile, polar reciprocal transformation; it gives a far wider range of interpretation of any given phenomena in the living world than does the one-sided and indeed old-fashioned approach to morphology, which relies only on Euclidean concepts.

For example, proceed, to begin with, quite simply, by allowing the point to draw a circle in the horizontal plane. To begin with it is easiest to take a circle symmetrically poised above the sphere, with its centre at the point where a vertical axis (not shown in Figs. 39 and 40) would strike through the horizontal plane. It will be easily recognized in imagination that the plane which is polar to the point drawing the circle will move in such a way that it will describe or plasticize a *cone of planes*, all passing through or held in the point-pole. The

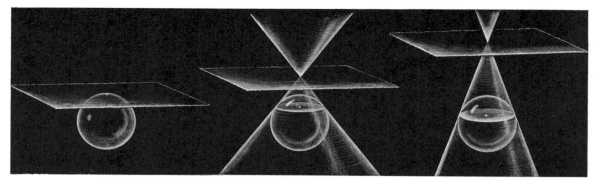

39   Polar Reciprocation between Centre and Infinite Periphery

40a   Pole and Polar

aperture of this cone will depend on the radius we have chosen for the circle in the horizontal plane, for this will determine the angle of inclination of the plane which is polar to the moving point (Fig. 41, Plate IV).

20 *"Extensive" and "Intensive" Two-dimensional Forms*

Our emphasis and terminology departs from what is customary in the textbooks of the new geometry, in that we give the fullest possible scope to the Principle of Duality and Polarity, and indeed express ourselves in such a way as to help to make the application of this principle a matter of instinctive feeling, as well as of pure thought. Thus we feel free to say that a plane is *in* a point, just as a point is *in* a plane; so we shall frankly speak of the *geometry in a point*, putting this over against *the geometry in a plane*.[17] We abandon the use of special terms – such as "sheaf" or "bundle" in the older textbooks – for the point considered as a manifold of planes and lines, but we introduce the idea of "extensive" forms in the extensive, two-dimensionality of the plane, and speak of the "intensive" forms in the intensive world of a point, which involves also the idea of the polar reciprocal aspect of two-dimensionality.[20] To some extent this manner of speaking has of course been used already, more in some textbooks than others. The departure is logically justified and indeed necessitated by the fundamental hypothesis of this book, which is to attribute to the idea of Polarity a universal significance for the spatial structure of the world, not only in pure thought but in the real activities of Nature.

Our circle in the horizontal plane, which we have learned to think of as formed by all its *points and all its lines*, is transformed by the sphere into a cone in a point, which we must think of as being formed by all its *planes and all its lines*. This is no ordinary cone in the sense of our familiar three-dimensional world; it is just as two-dimensional as is the circle, but in a polar reciprocal sense. We must learn to picture this cone as being swept out, or moulded from without by a moving plane, the plane which is polar to the point which draws the circle; it is a surface-form in the *"intensive"* world of the point-pole, which sustains it (Fig. 41).

No matter what form or figure may be described in the extensive realm of a plane, the sphere will call forth a polar reciprocal form – a conjugal form – in the intensive realm of a point. To comprehend such forms quite exactly, we must, as it were, allow thought to take wing and to reveal to the inner light of imagination forms which do not fully exist in the familiar earth-world of three dimensions. Indeed, all forms in this familiar world take on a new light, for ideally, according to this new way of thinking, to any form considered pointwise, there is a planewise aspect.

Furthermore, not only does the sphere or spheroid call forth, for any form in a plane, its reciprocal form in a point, but every continuous *surface in space*, formed as it is of points and of planes, will have its polar counterpart in another surface which it is possible to find accurately in "intensive" space. All forms are a

41   Cones in a Point and Circles in a Plane (Plate IV)

42   Spiral Cone in a Point (Plate VI)

synthesis of the two aspects, though the one or the other aspect might be dominant. Think of a surface as formed by a point that moves according to some determined law; assign, on the other hand, an appropriate law of movement to a plane in space, and the plane in its determined movement will mould and plasticize, or as mathematicians generally say, it will "envelop" a surface. These two polar aspects of form are related and organically joined by the sphere, which is itself a synthesis of the two. Just as the sphere has its pointwise, filled aspect *and* its planewise, hollowed out aspect (p. 55) so, too, have all forms. The cone of planes in a point, for instance, is enveloped by planes from without, leaving the "inside" hollow. Thus, the convexity and concavity of any surface will take on for us a deeper meaning in the new science of morphology, as will also the concept of "expansion and contraction".

21   Expansion and Contraction as a Qualitative Process

Let us go a step further, and picture for a moment that we have drawn not only one circle, but a whole family of concentric circles in the plane, growing outward from the central point in some kind of measure towards the infinitely distant line of the plane, into which at long last the largest circle will tend. Then surely the polar form will be a family of cones, rather than just one cone; they will also be one inside the other, but in a quite different way from the family of circles. We see such a family of cones depicted in Fig 41 (Plate IV). In this plate, the horizontal plane among the cones is tangent to a sphere in the same way as in Fig. 39 (left), but in the plate the sphere is not drawn in. This plane is therefore polar to the central point of the concentric circles, a few of which are drawn in. Looking back at Fig. 40a, it is not difficult to see that the points of a smaller circle in the horizontal plane will give rise to polar planes inclined more obliquely, and that the points of larger circles will have polar planes more vertically inclined, so that smaller and smaller circles, finally sharpening into the central point, will relate, in the polar forms of the cones, to wider and ever wider open cones, until in the last resort, when the circles will have shrunk to a point, the cones will have flattened into a plane (the plane which is polar to the central point of the circles). On the other hand, following the circles in the plane outward towards their infinitude, which is the infinitely distant line of the horizontal plane, we shall see that the cones close in towards a smaller and smaller aperture, towards what for them is an *infinitely innermost line* – the vertical axis of the picture. As the points of the circles merge into the infinitely distant line, the planes of the cones merge into the innermost axis.

This significant thought-form may be stated as follows:

Of the circles in the polar plane :
When infinitely large, their points are in the infinitely distant line of the plane, while their lines all become one with it. (This infinitely distant line is doubly covered by all the points and lines of the circle.) When infinitely small, their lines all lie in their mid-point, while all their points become one with it.

Of the cones in the point-pole :
When infinitely closed, their planes are in the inner (vertical) line of the point, while their lines all become one with it. (This innermost line is doubly covered by all the planes and lines of the cone.) When infinitely open, their lines all lie in their "median" plane, while all their planes become one with it.

Just as the circles have a central point from which they expand out into the infinitudes of the extensive realm of the plane in which they lie, so the cones, which are held in the point and open out (upward and downward), have a "central" plane from which they begin to close in towards their common axis, which is for them an innermost infinitude. We use here the word *median* plane, pairing it with the mid-point of the circles and thus differentiating between two kinds of "middle".[17]

It is a valuable and for our purposes indispensable exercise to familiarize oneself with this concept, to practise it in many possible variations and experience its polar qualities. In Fig. 40 (Plate VI), for example, the spiral surface would arise in the cone-point, if in the horizontal plane a point were to draw a logarithmic spiral.

Expansion and contraction now appear in a new light, or at least, the whole question may be thrown open. Quite evidently, expanding and contracting circles or spheres picture the idea of expansion and contraction, but what of the polar reciprocal process? When circles, beginning in a point, expand to the infinite periphery, in the polar space of the cones, the process starts from the median plane and proceeds towards the *inner infinitude of the vertical axis.* Is this to be thought of as expansion or contraction? And what is happening in the reverse process, when the circles shrink to a central point and the cones die back into their median plane?

The innermost point, which functions as an infinitude within (p. 57), will play an important part in our further studies. The nineteenth-century mathematicians had a good intuition, when they called it a "star". The sphere, as we have said, has two aspects; considered spatially, it has a central point and when it expands it tends towards the plane at infinity of space. Considered in its polar aspect, as an assemblage of planes, it has its plane at infinity, and the other extreme, towards which it tends in the polar reciprocal process, is its "star" – the point at infinity, or functional infinitude, within it.

In the great advance of science from the time of Newton onward, it is an essential feature that *ideal entities* contain the key to the understanding and mastery of phenomena in Nature in which they are only approximately realized. Indeed, they owe their power to their ideal purity and clarity – the very feature which makes them unrealizable to outer senses. For example, in mechanics, material particles can at most approximate to the character of a mathematical point, and yet this concept is essential to that of a centre of gravity, and without this and others like it we could not master the science. A scientific theory must not only allow for this fact; it must be founded on it. Truth is, the ideal entities are not approximate mental copies of sense-perceptible objects; they are the ideal aspect of a reality which presents itself to us only partially by sense-perception. Therefore, till joined to this ideal aspect, the sense-perceptible object remains dumb and inarticulate; united with it, it unfolds in full reality and only now we "know" it.

Goethe's idea of "expansion and contraction", so fundamental to his idea of the "archetypal leaf", is in effect an expression of polarity between a most contracted and a most expanded entity *within the space in question*. To entertain this idea not only in the aspect of the finite, already created, physical forms in space, but also according to the form-*creative* process underlying it, brings us close to Goethe's thought. The growing plant brings forth manifold shapes and geometrical patterns, on which the botanist bases his recognition of plant family and type. The plant reveals itself, now in expanded planes, now in hidden points. The living geometry and dynamics which Nature here unfolds belong to the more original and mobile projective and polar forms of space, with its balanced and *reciprocal polarity* of plane and point – expansive and contractive tendencies. It recognizes the shapes and patterns on a deeper level than the merely external and spatially sense-perceptible level. Concepts of size and quantity in outer measurement take second place, or at least refer better to end-results rather than beginnings. So it is with leaf and bud or eye; the expanded leaf-form has grown out into space and is finished and visible; it is, however, only complete when taken together with its bud or eye, the "intensive" form which harbours the future.

Such are the questions concerning expansion and contraction, which will interest us in later chapters of this book. We shall need to know and understand the ideal entities *point*, *line and plane*, each in their *intensive* as well as their familiar *extensive* character or quality. This is a conception as yet unfamiliar to experimental science in general; it has, however, long been close at hand in modern mathematics. The intense activity of pictorial but sense-free thinking here required leads beyond the boundaries determined by a study of morphology, which takes only into account the already created and sense-perceptible phenomena.

22   *Continuous Transformation and Polarity – Metamorphosis*

In all that we have been considering since entertaining the idea of the interplay of points, lines and planes at the beginning of this chapter, we have gone far beyond the fixity and rigidity of Euclidean thought-forms. At first the new element to enter in has been that of forms in *movement*. We have contemplated the transformations of forms one into the other by perspective, as for instance in the change from circle to ellipse and on into parabola and hyperbola. These are all sister forms within one family, and to pass from one form to the next is to be involved in a continuous movement and variation of shape. (The same process can be applied in the realm of the cone-type forms in a point and also of spherical-type forms in space.) This has only been possible through the acceptance of the idea of the infinitely distant elements, and in so doing, we have advanced from the idea of the discrete and separate forms of Euclidean geometry to the idea of *continuity* and *synthesis*, in which the whole process is primary and the parts are secondary.

The inherent polarity in the idea of point, line and plane and their inter-relationships, has, however, also led us to the far more radical, discontinuous and sudden transformations in the idea of the *Polarity called forth by the Sphere*. Movement and Polarity are as it were the basic ingredients of transformation and metamorphosis. To progress from fixed and rigid forms to forms in movement and then to polar transformations is a very useful geometrical exercise, undertaken by the mathematicians in recent centuries, which goes hand in hand with a study of morphology. It leads to a clearer and deeper understanding of metamorphosis than is at all possible with the limited and finite concepts of form and space, based on Euclidean geometry or even on the one-sidedly pointwise thought-forms of modern, "non-Euclidean" (analytical) mathematics.

We have been introduced to the possibility of a world of moving, changing forms, in great variety, in which the perception of an all-prevailing polarity of spatial structure is fundamental. Each form (curve or surface) has many manifestations; as we have said, we are not even limited in the polarizing process to the fixed form of the transforming sphere, for, like the circle, it has its sister forms.[21]

Goethe's idea of metamorphosis is indeed inspired out of the same realm of spiritual activity as is the idea of the polar transformation of forms.

In the fifteenth century, man had reached a stage in history in which he was developing an intense inner activity of thinking and at the same time was awakening in a new way to the realm of outer phenomena. The era of modern science was dawning. The overcoming of the limitations of Euclidean mathematics, as we have seen, brought the light and mobility of perspective into men's manner of observation in the sense-world. In many different ways, the universe was opened up for exploration. The direction taken was into the realm of physical and material exploration. Science was armed with the analytical geometry of Descartes, which, though freed from the fixed framework of the right-angle and the idea of rigid measurement, led, nevertheless, in the first place, to the one-sided "exact science" of our time, which has abandoned itself to the material, atomistic, pointwise realm.

The next stage, which was reached around the turn of the nineteenth century, has hardly yet been noticed; it is in the nature of things that processes set going in some direction tend to run on in the old lines as if by dint of inertia. This would seem to be a law in the process of evolution. Looking back into history and even into our present time, the origins of a particular development usually lie far back in the past, half hidden in the debris of opinions and theories, like half-buried seeds. This is as true of science as it is of Christianity.

The work of Goethe as a scientist is beginning to trickle into the science of today, notably in botany and the theory of colour. Rudolf Steiner, introducing Goethe's writings on organic morphology, calls Goethe the Kepler and Copernicus of the organic world, in that he has laid the theoretical foundations and established the methods for the study of *organic* nature.[22] The great difference between the phenomena of inorganic and organic nature is that in the former, the sense-perceptible occurrences are determined by conditions which likewise belong to the sense world. Here the concept and the phenomena coincide. In the organic world, however, sensible qualities appear, the causes of which are not immediately clearly perceptible to the senses. In this realm it is not easy to grasp the unity of concept and percept – to perceive exactly the truth underlying the observed phenomena. It is in this realm that Goethe's great contribution to science lies.

In the organic realm, full understanding of the whole phenomenon is not to be found in a study of the finished, material form alone, but only when the spiritual idea or concept which belongs to it is joined with the percept; in the inorganic realm the material form or physical process itself furnishes the required data. Compare this with what we have been considering in terms of the development of mathematics. In Euclidean terms, the comparison, for example, of similar triangles rests with their actual, observable measurements, just as an inorganic form is adequately described in terms of its observable measurements and material constituents. In the examples afforded by modern geometry, such as the cube and the octahedron in their intimate relationship with one another through the law of polar reciprocal transformation, the comparison must be undertaken in quite a different way. The difference between cube and octahedron is very great, until one enters into the *idea* whereby they are related. In the theorems of modern projective geometry, the statements are *pure ideas*, which are then revealed in manifold external ways, as soon as one puts them into practice in pictorial imagination, or reveals them in a drawing. The idea hovers among the many possible manifestations and their changes from one form to another, just as Goethe's *Idea* of an archetypal plant hovers above all individual plants and plant species.[1] Here we touch the nerve of what was being envisaged as a future direction of science by Goethe and many of the seekers of his time, and the great mathematicians did not lag behind the task, even though their efforts were destined to be lost sight of for a time.

23   Goethe and Modern Geometry

The morphological ideas derived from modern geometry give a basis for a systematic training in "metamorphic" thinking. Goethe did not reflect much on his own theory of knowledge, yet a true method and consistent theory of knowledge underlay his work. He believed that *the Theory is contained in the Phenomenon*. The phenomena our senses see are not unlike the ideal reality which brings them forth. As the Greek origin of the word implies, the Theory is the true seeing of the thing – the insight that should come with healthy sight. Yet man is so constituted that he does not really see, unless he meets what he sees with spiritual activity on his own part. It becomes the role of thought to interpret the language of phenomena. If what our thinking reveals is *in* the phenomena the senses see, so that, enlightened by thought, we feel that we are seeing with new-awakened eyes, then will our thinking have been true; it will have met and united with the archetypal Thought whereby the phenomenon was created. Scientific thought should not lead away from the phenomena into some remote construction from which the thing perceived derives by elaborate causation. It must give us the faculty – latent but unawakened in the merely passive mind – to read in the phenomena, so that the chaos of detail becomes articulate language.

Goethe's awakened faculties were expressed in his whole scientific outlook. It may at first cause surprise to claim that there is an affinity between the modern geometrical approach to form and Goethe's work. Yet as some modern authors have recognized, there was far more of the mathematical in Goethe's thinking than might at first appear. He was far too universal a spirit to have allowed the new mathematics, which was coming into being in his time, to pass him by. But mathematics, too, has gone a long way since Goethe's time, and we are nearer today to an appreciation of the qualitative and form-creative as against the merely quantitative aspect of this science. It is this qualitative aspect which reveals the air in which Goethe breathes.

The "holistic" strain in Goethe's thinking allows him to perceive the organic relation between part and whole, or between part and part. He begins by observing the whole, – the whole which contains the parts, and then he sees that in each part the virtue of the whole is contained. For Goethe, this became an immediate aesthetic insight, and so it is for many a biologist in gifted moments. The concept of an ideal Type, having many manifestations, outwardly often very different from one another, perhaps even unrecognizable, we have seen revealed in the few geometrical studies we have undertaken in this chapter. We have seen it, for example, in the family of curves in Fig. 22, where the law of the whole is to be found in each separate curve, however different they may be from one another externally; we have seen it in the examples of polar reciprocation, where the ideal type reappears transformed beyond all external recognition.

In a still deeper sense the change of outlook in science is contained in the later work of Rudolf Steiner, who relates mathematics and imagination rather in the spirit of Goethe's essay on "Perceptive Judgement".[23] In selfless contemplation of Nature, says Rudolf Steiner in effect, we can so train our imaginative faculties that they become an instrument of cognition no less conscious than mathematical reasoning. Our mathematical thought-forms – such as the three perpendicular axes of analytical geometry – spring from the form of the human body and from the upright stature we acquired when in infancy we learned to stand and walk. All our cognitional faculties in earthly life are in some way a sublimation of the powers of life and growth which placed us bodily into the world in embryo-life and early childhood. Many-sided is our common bond with Nature; only by virtue of it can we know her.

Euclidean thought is of such a nature that it supplies the concepts which fit the world of *created* forms, which may be the object of our attention in the external world of space at any one moment in time. Trivially speaking, as a material body, the plant form or any organic form is there, of course, in three-dimensional space. The question is, whether that is the whole story, if one would understand what has taken place before that moment, and what will eventually be determined through the process of Becoming. The organism lives through sequences of time; the changing forms of the plant arise even as colours arise in the interplay of light and darkness. Here too, it is a matter of seeing the whole and not only the part; colours arise with their different qualities and shades *between* the polarities of light and darkness. In "Light and Darkness" Goethe divined an ideal polarity, fundamental to the structure and processes of the world; the visible phenomena were for him an outcome of the interplay of both components.[24] In these realms live the thought-forms of projective morphology.

24   *Concerning Conceptions of Space*

Our considerations so far involve another outlook into the spatial universe as such. The cosmic outlook in the age of science hitherto has been instinctively and naturally pointwise: problems connected with the seeming infinitude of the universe, with the nexus of celestial phenomena and the place therein of the solar system and the earth, have always started from this premiss. If now the basic polarity of space – that of point and plane – is to have meaning for the real universe, there will also be an aspect of cosmic and earthly nature as a whole, polar opposite to what is pointwise. We have to look for this aspect and find for it an appropriate designation.

Here, too, the history and development of modern geometry points the way. It was discovered by Arthur Cayley and Felix Klein, during the second half of the nineteenth century, that the more specialized and rigid metrical forms of geometry could all be based on the more general and mobile projective form. In this way the ordinary space of Euclid – answering to our everyday experience of the physical world – and also the different non-Euclidean spaces could be brought together. It was only necessary to propound a unique entity – as it were, cosmically and absolutely given – whereby the definite measures of a particular space should be determined. They call this entity "the Absolute". For Euclidean geometry it is an infinitely distant plane (with an "imaginary circle" therein inscribed).[25]

Mathematicians are used to many different conceptions of space, – Einstein's space among them. Around the middle of the nineteenth century, when the discipline of projective geometry was already well-established, there lived in

Germany Hermann Grassmann,[26] a man of repute in two widely differing realms of culture; he was a scholar in old Indian manuscripts and in mathematics. His most significant work remained long unnoticed; according to A. N. Whitehead, he was "a hundred years ahead of his time".

Grassmann proved, following the direction of projective thinking further, that a polarity of contracted and expanded entities, analogous to point and plane, will hold good in spaces of any number of dimensions. He founded in this sense what he described as a "science of extension" (*Ausdehnungslehre*). We shall contend that in the relatively simple threefold space of our universe, the living plant is like a realization of such a polar "science of extension". We shall attempt to recognize the plant as a being in Nature which *lives and makes manifest the processes of its becoming*. Unlike the material, finally created, dead world of the mineral, which lies as a finished object in a fixed space, the plant, *in becoming*, forms its own space, and it does so according to the same ideal laws which underlie the creation of universal space; laws – or ideas – which can light up in the inner perception of man in the activity of thinking. In the three-dimensional space of our universe, the plant lives out, as it were, in the real world a "science of extension".

In this spirit we shall attempt to bring together with phenomena in the plant world, and in particular in regard to the leaf, the idea of a plane as an original, form-creating entity, equal in importance to the point. We will put to the test, without bias, what arises when the world of phenomena is approached with the following direction of thought. Just as in the condensed and rounded off material body – a stone, for example – the idea of the point (as centre of gravity and so on) applies, so in the leaf, revealer of all plant nature, it is the plane. If this be true, it will show how point, line and plane have their counterpart in the plant, namely, the line very obviously in the stem, the point in the seed or eye, and the ideal plane in the leaf.

This co-ordination, however, will become meaningful only if we deepen our conception of "point" and "plane", as we have begun to do, seeing the plane not only extensively, as the sum of all its points, but also intensively, as a single whole; likewise the point, not lying dead as a speck of dust, but alive, for example as a moving organism of planes. On the accustomed level of Euclid the comparison would, of course, be trivial.

Above all, we shall need to understand how the leaf is Nature's adumbration of the ideal plane. Then, too, we shall attach fresh significance to the universal association of leaf and eye – the organ that shows forth the archetypal form with the one that only bears it for a potential future. Goethe perceived that the expanded, planar organ has its inseparable counterpart in the contracted. In a letter to Herder on May 17th, 1787, he wrote: "It dawned upon me that in the organ we are wont to call a leaf the Proteus lies hidden, who can reveal or veil his presence in all the forms the plant brings forth. Forward and backward, the plant is ever leaf, inseparably united with the germ of the future plant, so that the one is unthinkable without the other."

The mathematicians, following Cayley, Klein and v. Staudt, call the "space" in which the balanced, creative interplay of polar forms takes place the (three-dimensional) "projective space". It is the "space" which presents itself to the activity of thinking and inner imagination, when one moves freely in the realm of the exact relationships between Point, Line and Plane, considered as manifolds or organisms, and once one recognizes that the ideal, infinitely distant points and lines, and even the infinitely distant plane function in just the same way as all the others. It is a realm attainable in thought, in which one works with continually *transformable* form-types, and not with ready-made, rigid figures. We call this mathematical thought-realm the *free Archetypal Space* (Freie Ur-raum).[17]

To be active in this realm of thinking, one sets aside, to begin with, any idea of measure, and one takes the infinitely distant elements for granted. In the weaving interplay of planes, lines and points in Archetypal Space there are myriad possibilities for the creation of all manner of forms – not only forms with straight edges and flat surfaces, but curves and curved surfaces also. It is a realm of infinite potential. Once a process has been begun, the planes and lines have the tendency to weave together according to some superior law; they will not be in disarray, but will create sensible patterns. Out of these patterns and spatial forms (one might call them "achieved relationships") measures then arise functionally; they are *not* predetermined. Thus, in contrast to the ancient geometry of Euclid, we do not begin with a measure, or measures, and then build up forms from these; it is just the other way round. If amid the moving, weaving, mutual interplay between point, line and plane, we establish some "Absolute" – some invariable element – then a measure will become manifest in some definite type of form or in a metrical space of some kind or other.

The rhythm of measure is thus freed from any preconceived fixity. In the Archetypal Space of projective geometry there *are* no rigid measures; there are only *functional relationships*, which find expression in pure number or form.

With this concept of archetypal space, the mathematicians reach into a realm of thought for which the words "space" and "geometry" are too narrow – even misleading; for as yet no actual metrical space, nor measurable forms have been created, and even when they come about, it is an exception rather than the rule to find them expressing the rigid measures of *geo*-metry. It is a realm in which the archetypes of all manner of spaces and forms are to be found. Perhaps one is here permitted a glimpse into eternity – into the time when God created the world, and man in his own image.

Out of an infinitely free potential of rhythmic and polar interplay between the three archetypal elements, the many metrical space-concepts and the geometries therein contained may be derived, by planting in the projective space an invariant Absolute, by relation to which the prevailing measures will arise.

III
71

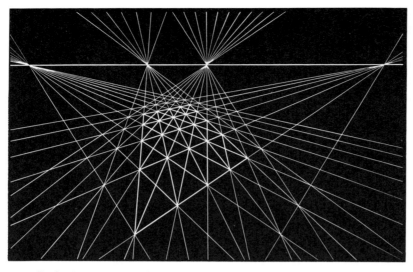

43 *Projective Net of Quadrangles in Step-Measure*

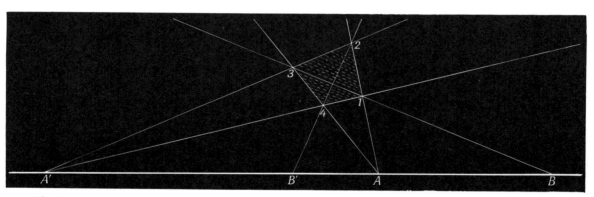

44 *The Harmonic Quadrangle*

45 *Regular Net of Quadrangles in Step-Measure*

In the so-called "projective determination of measure", there arise to begin with three basic types: hyperbolic, parabolic and elliptic. We, using simpler language, will call them Growth-measure, Step-measure and Circling-measure. For the purposes of this book, we will approach the ideas concerning the different measures as simply as possible, and will refer the reader to the Notes and References and to other writings for a fuller and more systematic approach.[27]

Step-Measure. For Euclidean space – that is for the familiar space of Earth, which we experience in everyday life – the Absolute is an infinitely distant plane, bearing an "imaginary circle".[25] Euclidean space is a three-dimensional projective space, in which this plane and circle are kept fixed. The resulting measures are parabolic, that is to say, of equal steps (arithmetical progression). In physical space, we take a ruler, set it down at some point, mark the point at the other end of the ruler, and then repeat the process again and again. Ideally, we could continue this repetitive process *ad infinitum*. The laws of equidistance, of parallelism and of translational movement are due to the infinitely distant plane, and the basic forms of the right-angle and of rotational, circular movement are due to the invariant imaginary circle.[25]

As we have learned, projective geometry treats as equivalent forms which are capable of mutual transformation by perspective, or by a sequence of perspectives, and does not regard the planes or lines at infinity as being different from any others.

In constructions of perspective, we put "vanishing points" on "vanishing lines", bringing into the actual picture what is otherwise ideal, namely, the infinitely distant line of the plane; this line represents the horizon from any assumed point on the plane. The same applies in the construction of plastic (three-dimensional) perspectives – as in the theatre, or in a bas-relief – where there will then be a "vanishing plane", representing the infinitely distant plane as such.

In the projective plane in Fig. 43, the chosen Absolute is the horizontal line, on which there are four points from which the lines ray forth to create a network of quadrangles side by side. Once one quadrangle has been freely drawn, as for example, in Fig. 44, all the others arise as though of their own accord. Indeed, one must realize that *once the weaving process has been set going, the whole net is already there*, throughout the whole plane – invisible, until it has been drawn! The lines weave throughout the whole plane, above the horizontal line as well as below, tending towards it, as in a perspective picture, no one quadrangle having the same measurements as any of the others. It is called a harmonic net; there is evidently a hidden law at work, which maintains order throughout the plane.[28]

Now bring movement into the picture: let a point move, say the one to the left on the horizontal line. Then all the others – the whole network – will move accordingly, changing to outer appearance, but always remaining true to the original idea. It will be interesting to try out what will happen, if one of the points moves to the infinite point of the horizontal line. Surprisingly enough, it will then be found that the net becomes more regular; a step-measure will appear on all the lines raying in from the point at infinity! Rigidity has begun to set in.

Fig. 45 shows what happens when the *whole* (horizontal) line has moved to the infinite. The Absolute now is the invisible, infinitely distant line of the plane, and the figure reveals the rigid step-measure, which results when the whole line, with its four points functioning as the Absolute in the projective picture, moves to the infinite! At this moment, the Absolute is the invisible line-at-infinity of the plane. From this it can be seen that the vanishing scale of measures in Fig. 43 is a picture of *step-measure in perspective*.

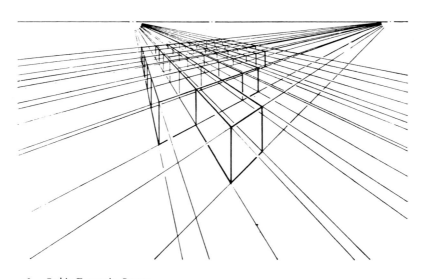

46   Cubic Forms in Space

Fig. 46 shows the similar process in three-dimensions; here a plane containing an archetypal triangle functions as the Absolute for a three-dimensional form. From the points and lines of the archetypal triangle, the lines and planes ray out, creating the three-dimensional form in three-dimensional projective space. It is in all respects, *apart from measure*, a cubical form (parallelopipedon).

Allow the archetypal plane to recede to infinity and, geometrically speaking, the elementary crystal lattice *is* such a harmonic network, formed from an invisible, archetypal, harmonic pattern in the plane-at-infinity of Euclidean space. The projective picture allows us to see what is really happening ideally, even in the actual parallel-sided form, such as a crystal.[29]

See the cubic forms one beside the other, as was done in the two-dimensional picture of quadrangles, and the ideal picture revealing the quality of Euclidean space becomes evident to the mind. It is a space capable of being entirely filled

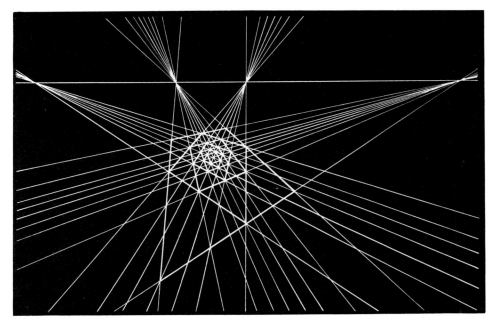

47   Projective Net of Quadrangles in Growth-Measure

48   Regular Net of Quadrangles in Growth-Measure

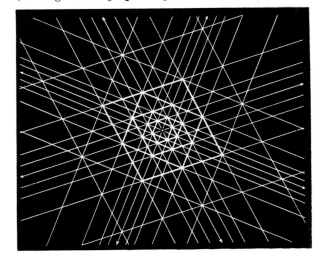

with forms *side by side*, stretching away to the infinite distances, with no particular centre — one might also say, where each single form can be regarded as the centre. *This space has many centres and only one infinitude, the archetypal plane.* For the cubical form the archetype is a triangle in the plane at infinity; for other crystal forms there will be other patterns in this archetypal plane.

Here is revealed the true nature of Earth-space, with its repetitive step-measure and its finite forms; here is the old Euclidean idea, but with the addition in thought of the infinitely distant elements, the existence of which Euclid denied. It is as though Euclidean space itself were a texture of woven light, or were shot through with countless textures of this kind, formed from the plane-at-infinity, and every crystal a partial embodiment of such a texture, rhythmicized in its own individual measures. In effect, the measures that prevail throughout the mineral kingdom and that dominate the crystal form with its internal lattice, derive in the purely geometrical sense from the infinitely distant plane of our familiar space. More than that, the crystal form itself is a projective construction, with its archetype in yonder plane.

This is the inner reason why crystallographers portray the crystal types by projection on to a surrounding sphere. The latter has no particular radius; it is in fact a finite image of the infinitely distant. Upon this finite sphere we draw an image of the infinitely distant archetype. To form the crystal, geometrically speaking, it is as though the archetype sent its rays inward. Potentially, it could fill the whole of space with its crystal lattice. Where there is mother-liquor to receive this in-forming, the potential form is materialized and made visible to outer sight.

The significance of these discoveries — essentially, they date from about a hundred years ago — can surely not escape unbiassed thinking about Nature. But it has been the fashion to suppose that all scientifically real thought-forms must be derived by abstraction from the properties of tangible, sense-perceptible objects. Hence, if a valuable thought-form, such as these infinitely distant or even imaginary elements, is obviously not of this kind, the tendency has been to regard it as a mere arbitrary definition or convention, – a mere word we find convenient to use in a fictitious sense.

This way of thinking was no doubt natural to the materialistic realism of the nineteenth century. Today there is every reason to go beyond it.

Growth-Measure. To make the transition in a simple manner from step- or parabolic measure to growth- or hyperbolic measure, all we need do is to bring in the idea of forms *one inside the other* (Figs. 47–49), instead of forms side by side. In so doing, the circle and the sphere are close at hand.

Take a square and inscribe within it, for example, a circle; or take a cube, as in Fig. 35 and inscribe in it a sphere. In the projective picture, freed from the rigidity of the right-angle and the element at infinity, the harmonic net will reveal forms one inside the other, as in Fig. 47 (in two dimensions) or in Fig. 49 (in three dimensions). Here the octahedron arises in relation to the cube.

The measure arising in this type of net will be of steps growing — or diminishing — in such a way that two adjacent steps will always show the same

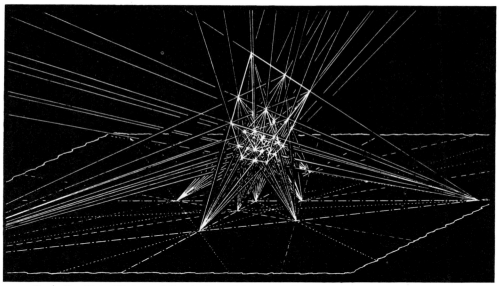

49　*Projective Net of Cube and Octahedron giving Growth-Measure*

proportion (geometrical progression). We call this "Growth-Measure". It will be seen that there are here *two infinitudes*, for the sequence progresses outward towards the infinitely distant line – or plane – and inwards towards a "star point" – a point *functioning* as an inner infinitude. In Fig. 47 the forms close in towards such a point within and open out, to flatten in the last resort from both sides into the horizontal line.

Once again, what we see here is a perspective picture of what in ordinary space would be a logarithmic process – a growth-measure taking place between an innermost point functioning as an infinitude, and the line or plane at infinity of space. In Fig. 49 the perspective picture does the same for a sequence of three-dimensional forms; cubes and octahedra would alternate, one within the other in polar fashion, between the innermost infinitude and the plane at infinity of space.

It is important to appreciate the difference between step-measure and growth-measure, in that in growth-measure we have to do, not only with one, but with two infinitudes; here the measure runs between the functional infinitude within and the actual infinitude without, the latter is the *same* as the one distant infinitude upon which step-measure depends.

Circling-Measure. The third basic measure is circling-measure, and this plays a part in the realms of both the other measures. Consider a family of concentric circles and their radii, as for example, in the top left-hand drawing in Plate X. Here there are two possibilities: either we may follow the growth-measure of the circles inward or outward, or we may go round and round in the measure given by the radii; this latter is a circling-measure of lines in a point (the point-at-infinity within), and of points on a line (the line-at-infinity without). In Fig. 50

we see the same circling-measure projectively portrayed along a visible line; were the whole family of curves to be drawn in, we should see a projective transformation of the concentric circles, which would look something like the curves in Fig. 22.

In circling-measure, the archetypal principle of the rounding off of spatial forms comes to expression. In its physical, spatial manifestation, it appears in its purest form in the plane as a circle, in the point as a circular cone and in space as a sphere. It is related to everything which circles or makes vortices in the universe, and also to the archetype of everything which is closed in upon itself and has come to rest.

Every living being tends in some way towards a rounded off and finished form. It is perhaps natural that the most primitive living forms tend in their visible spatial body towards the sphere.

This principle, too, arises to begin with out of the free Archetypal Space and not in the rigidity of Euclidean measure. Between the extremes of expansion and contraction, together with the circling principle, all forms come into being. Unite the growth-measure with the circling-measure, and those forms arise which are most ideally balanced between the two principles, namely the spirals. On the basis of the growth-measure in Fig. 48 the spirals as in Fig. 51a arise, where the radial as well as the circling component is brought into play. There arise spirals, which when drawn out in their continuous form are to be seen in Fig. 51b. According as to whether the radial or the circular component dominates, the spirals will be less or more extremely curved. These spirals run between the two infinitudes: the innermost, which they never reach, and the outermost line-at-infinity which is also physically unattainable. Based on growth measure, these spirals are called logarithmic, or, as they are often called in a more metaphysical sense, the "Spirals of Life".[30]

Spirals do arise also in the interplay of step-measure with circling-measure. Take, for example, a family of concentric circles, getting larger in step-measure from a central point. Here the circles, together with their radii give rise to spirals which have no inner infinitude. These are the so-called Archimedian spirals, which run from a finite point out to the infinite and are qualitatively very different from the logarithmic spirals.

The growth-measure spiral, or spiral of life, has long been associated with living forms, especially of the plant world; it comes to expression also in water, upon which all life depends. It represents the rhythm of the nodes up and down the stem of the plant, in the spiral phyllotaxis. We see it pictured again and again in Nature, most often in the growing-point of a plant, as in Fig. 52, where it is always to be found, as though telling of the origin of the tiny leaf primordia. We see it frozen into the forms of shells, left behind like a record of its life by the watery organism (Fig. 53).

The difference between the Spiral of Archimedes and the Spiral of Life is fundamental: the former with its one infinitude and its repetitive step-measure (addition), the latter, as in Fig. 51, spanned between two infinitudes, with its proportional measure (multiplication). Circling-measure belongs to both

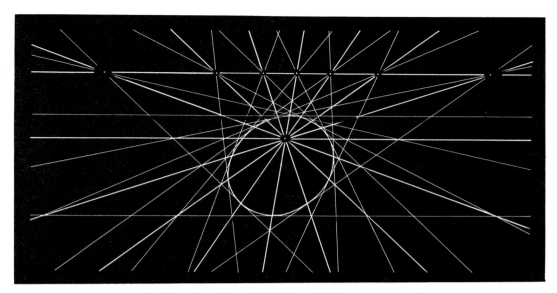

50   *Circling-Measure in the Lines of a Point and the Points of a Line*

51   *(a and b) Logarithmic Spirals (Growth-Measure)*

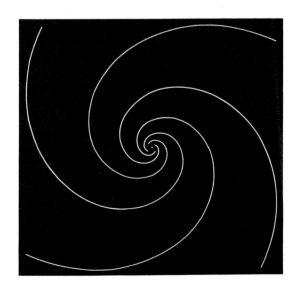

52   *Leaf-primordia at the Growing-Point (from Church)*   53   *Section of Nautilus pompilius*

Spiral Forms in Nature

54   *Sunflower*   55   *Italian Cauliflower (from Grohmann)*

realms; both the Archimedian spiral and the logarithmic spiral circle round and round, inward and outward. In the former, however, there is a "dead end" in the centre, while in the Spiral of Life the curve continues on and on inward, just as it goes on and on spiralling outward. It spirals in towards a point, which has the same *quality* as the infinite periphery towards which it spirals out; this point plays the part of an innermost infinitude. We have called such points "functional infinitudes", because, although in ordinary space, they function as though at infinity.

Unlike the inorganic forms, all living forms have an inner realm as well as the outer one, and they live their life according to rhythms, which play between the two. We touch here upon the difference between "beat" and "rhythm". *Beat* is the mere repetition of a sound at equal intervals, while *rhythm* is spanned between infinitudes.

The logarithmic spiral is beautifully portrayed in the distribution of florets in a composite flower, such as the Sunflower in Fig. 54. It is pictured in a more spatial form in the Italian Cauliflower in Fig. 55. It is to be seen in pine-cones, in the horns of animals, in bones, in muscular organs, such as the heart, and in the forms made by water, air and warmth. Indeed, this curve and others akin to it are dear to Nature's heart. (See § 46 and Note 64.)

Chapter IV

PHYSICAL AND ETHEREAL SPACES

27 *Space and Counterspace*[13, 14, 15]

From the deeper concepts of space to which projective geometry has led, we derive a theory concerning the type of *formative field* which the specific phenomena of life reveal. The theory will obviously bear on all living forms – plant, animal and man. We introduce it in connection with the higher plants (the cormophytes and above all the flowering plants), since it is here that the type of form and unfolding growth which the geometrical ideas suggest is most directly evident.

In the foregoing chapters we have stated it descriptively, bringing in the geometrical ideas with the help of simple illustrations and in imaginative language. We shall now characterize more precisely the scientific notions employed and draw attention to the problems they involve, addressing ourselves at this point to scientific readers. The reader untrained in mathematics need never lose heart, for just as one grasps for practical purposes many a realm of modern technical science, without necessarily following in detail all the underlying mathematics, so, too, in this realm of the ethereal forms and forces.

The path along which we are led is as follows. Projective geometry as we have seen treats, as equivalent, forms which are capable of mutual transformation by perspective or by a sequence of perspectives. It studies, therefore, mobile *types of form*, rather than rigid forms. Moreover the laws differentiating one type from another spring from those interweaving relationships among the planes, lines and points of space which are made use of in all perspective constructions. This has a twofold outcome, which proves of great importance when we begin to see the real universe of Nature in this aspect. It makes us far more aware of the totality, and above all, the infinite distances of space. In the constructions of perspective the latter are not left as a vague and inconceivable infinitude as in the ancient geometry of Euclid. They are transformed into quite tangible "vanishing points", "vanishing lines" and so on, thus making manifest the real part they play in the spatial structure of the most material and finite forms – the crystals, for example.

The second outcome is that projective geometry overcomes the one-sidedly pointwise idea of three-dimensional space which prevails in the old geometry and, to a great extent, in our instinctive feeling. It reveals the interplay of two mutually polar and complementary principles – pointwise and planar, or centric and peripheral. The Principle of Duality or Polarity, consistently applied, shows that the plane plays a no less primary and fundamental part than the point in the formation of space.[17] Weaving between the two, with a perfectly balanced relation either way, is the third fundamental entity, the straight line. All this is of untold significance for an imaginative and at the same time reasoned scientific understanding of the world of Nature.

Attention has been drawn to the evident relation of plane and point to Goethe's ideal conception of expansion and contraction. The plane is what the simplest of spatial forms, the sphere for example, becomes when expanded without limit; the point, what it becomes when contracted. It may well prove that Goethe, both as botanist and physicist, divined a truth, the full and clear expression of which belongs to the science of the next hundred years from now.

Projective geometry has also led to a clear understanding of how and why it is that the perfect polarity of space as regards point and plane is not immediately manifest in the more rigid space of Euclid, and was therefore unknown until a little more than a century ago. It is due to the decisive part played in the space of Euclid – the space in which rigid bodies, for example, preserve their shape and size as they move – by the infinite periphery, which has the character of an unique, infinitely distant plane, the "plane-at-infinity" of space. This cosmically given plane determines the laws of parallelism which play so great a part in the translational and sliding movements of rigid matter; it determines the space of created things.

Within the infinitely distant plane there is also the invariant imaginary circle, to which the basic forms of right angle and of rotational, circular movement are due,[25] even as the laws of parallelism and of translational movement are due to the infinitely distant plane as such.

Euclidean space is a three-dimensional projective space in which the plane-at-infinity, bearing the imaginary circle, is kept fixed. The presence of an unique plane, with no unique point to balance it, upsets the all-prevailing polarity of point and plane, which is only restored when by projective transformations one, as it were, unfreezes the "Absolute" and lets it move.

A most revealing paradox emerges, of which the philosophical implications are far-reaching. The polarities of spatial structure are so interwoven that the very realm in which *point*-centred entities (atoms, material bodies with their centres of gravity, electric and magnetic poles and so on) are at home is a realm determined by an unique, cosmically given *plane*, which in the nature of the case is inaccessible to any material body – and, no doubt some would say, inconceivable to any mind inhabiting such a body! An unique entity of the expansive kind holds the field for the relationships, both geometrical and dynamical, of multitudes of contracted, pointlike entities. Such is the space of Euclid, which we may call *physical space*, since the typical laws governing the behaviour of inert matter – the laws of elementary physics and mechanics – are very intimately related to the structure of this space. (The relationship is brought out in the parallelogram of forces, the law of moments, in vector analysis and other geometrical devices by which one estimates the interaction of physical forces.)

The same versatility and detachment of modern thought which enabled projective geometry first to liberate itself from the rigid laws of Euclidean space and then, returning to it, to understand this space in a deeper aspect, gave rise to the idea of other forms of space, subject like that of Euclid to the more general projective laws and yet of different spatial structure. Among these are the "non-Euclidean geometries", which have been partly applied in the Theory of Relativity. Discovered in the early nineteenth century by Lobachevsky, Bolyai, Riemann and others, their inclusion in the wider framework of projective geometry was only afterwards made clear through the work of Arthur Cayley and Felix Klein. Now it is also possible within this wider framework to conceive a type of space – the precise "dual" or polar opposite of that of Euclid – which, we shall try to show, belongs essentially to the phenomena of the living world.[14, 15]

Euclidean space being a projective space governed by the invariance of an unique, infinitely distant *plane*, the polar opposite type of space will be a space determined by an unique *point*. The latter too will have the functions of an infinitude, but this need not mean "infinitely far away" in the ordinary sense, since to apply such a criterion would be to think again in terms of physical space. On the contrary, whereas the infinite periphery of the latter is in the nature of the case an *infinitude without*, the unique point of the polar opposite type of space will have the character of an *infinitude within*. We shall be likely to find it by looking not outward but inward, into the innermost heart and core of the spatial field in question. To use the very suggestive terms coined by Dr. E. Lehrs[31] in his book *Man or Matter*, whereas the space of Euclid is governed by an *all-embracing plane*, this other space is governed, as it were from the very innermost by an *all-relating point*.

The space determined by an "all-relating point" being polar to that of Euclid in all respects, its elementary entities (of no dimension and of zero volume) will be planes and not points. Its finite forms will tend to *envelop* the point-at-infinity within, and volumes – or substantial content, if such there be – will be calculated from without inward. The top right-hand drawing in Plate X for example, pictures a family of spheres in this other space, growing in volume from without inward. *The finite "planar volume" of each sphere is the entire space outside the surface; the remaining hollow, towards the point-at-infinity at the centre, though physically finite, is infinite in the measures of this other space.* The latter is indeed the true negative of Euclidean space – related to it, in a qualitative sense, as the mould is to the cast – and we may therefore call it "negative-Euclidean space", or even *negative space* pure and simple.

For the full determination of a negative-Euclidean polar space, the point-at-infinity must be imagined as bearing an absolute imaginary cone, polar to the absolute imaginary circle in the plane-at-infinity. To begin with, we take this cone to be "spherical"[20] – in direct perspective with the absolute imaginary circle. This assumption is, however, not essential; the absolute cone may be elongated or flattened ellipsoidally in one direction or another.

28 Relation to the Non-Euclidean Spaces

The significance of our hypothesis – namely that this other kind of space plays its part in the structure of the real universe – is brought out very clearly if we take our start not only from Euclidean space but from the two kinds of non-Euclidean space, all of which emerge from the more universal form of projective space as shown by the Cayley-Klein theory. It is well known that Euclidean or "parabolic" space is a degenerate transition-stage between the two non-Euclidean types, known respectively as hyperbolic and elliptic.[32] These spaces are obtained if the "Absolute" is supposed to be a *finite* sphere or quadric surface, of real or imaginary radius, say a, which is transformed into the (doubly covered) plane-at-infinity when a grows infinite.

So long as a is finite, the principle of duality is preserved, though in different ways in the hyperbolic and elliptic cases. For in the former instance the absolute sphere, being real, separates off the interior from the space outside. The interior will naturally be regarded as a pointwise space, and it was this alone which geometricians had in mind when these ideas were first developed. Within this space, it is true, the perfect polarity of point and plane breaks down, but the remaining space outside, *regarded as a planar space*, is polar to it with respect to the Absolute, and the two spaces, taken together in their mutual relation, do full justice to the principle of duality. (A. N. Whitehead, in his early work *Universal Algebra*, calls the space outside the Absolute the "anti-space", corresponding to the "space" within.)

If the Absolute is imaginary (elliptic space), there is no real surface to sunder the predominantly pointwise from the predominantly planewise realm. Space remains one and undivided, and to this single space the principle of polarity applies in its full scope. The nineteenth century mathematician and philosopher W. K. Clifford,[33] lamenting the fact that the balanced harmony of the Principle of Duality no longer applies in the seemingly Euclidean geometry of the real world, suggested that universal space was perhaps after all elliptic, but with a finite value of a so large as to have escaped detection. "Upon this supposition", he writes, "the whole of geometry is far more complete and interesting; the principle of duality, instead of half breaking down over metrical relations, applies to all propositions without exception."

We may now picture the continuous transitions between these different forms, letting a vary throughout the range of real numbers. Taking our start from the hyperbolic form, with a real absolute surface dividing space from anti-space, if we increase the value of a towards infinity the surface expands; the planar anti-space outside is reduced, and when at last the Absolute expands into a plane, the anti-space is flattened into nothing. Meanwhile the pointwise space inside, from being hyperbolic, becomes parabolic, in other words Euclidean. It becomes "physical space" as we know it.

If on the other hand, beginning again with hyperbolic space and anti-space, we *decrease a* (or, on the planar side, increase a^{-1}), the absolute sphere will draw in towards the centre. This time the pointwise space inside will be reduced, whereas the planar space outside grows inward. When at long last *a* becomes zero, it is the former space which is reduced to nothing; only the planewise space outside is left. This in its turn becomes negatively parabolic – inside-out Euclidean, one might also call it – oriented towards and Absolute which is now the point-at-infinity within, bearing an imaginary cone.

There are in fact two possible transitions from real to imaginary finite values of *a*, namely through $\pm\infty$ and o. The former leads to the plane-at-infinity and so to Euclidean space with its pointwise bias. This one alone was given serious attention hitherto. The other leads to the negative or polar counterpart, – to a parabolic space of which the primary elements are planar, determined by a point-at-infinity in the innermost. This would suggest that the polar symmetry of which Clifford felt the lack is to be looked for, not in an universal space given once for all, but in another way; namely that the Euclidean space of the physical universe, single and relatively everlasting – co-extensive in time with this universe itself in its present form – is pervaded by a myriad relatively transient negative space-formations. These, in their spatial character, represent the missing counterpart of the space of Euclid. This is the space referred to in so many ways by Rudolf Steiner; the idea of a kind of space so essential for the understanding of the laws of the living kingdoms. He called this space by many names, – "Negative Space", "Ethereal Space", "Sun-space", and "Gegenraum".

29   *The Essence of the Seed*

Wherever in the realm of living things there is a seed or germinating centre of fresh life, there is the "inward infinitude" of such a space. The ideal relation to ordinary space is a true polarity; only with respect to time, and with respect to unity and multiplicity, the two kinds of space play quite different parts, thus making possible the side-by-side and transitory existence of multitudes of living creatures.

Physical space is there for all time – at least as long as Earth-evolution lasts, and in this space, time is measured spatially. The hands of the clock or the shadow on the sundial mark off equal distances or equal angles during the passage of the hours. Even this, as the leap-year tells, is a compromise, for the Universe itself is alive, and not bound and limited to a rigid *geo*-metry!

The ethereal spaces, on the other hand, could really be called "Time-spaces"; they come and go in the interplay of the cosmic and earthly rhythms. The seeds lie dormant in Earth-space, until such time as the cosmic processes of the space of which each one is the inward infinitude begin to be active, in harmony with the cosmic rhythms of the seasons around the Earth. Then the plant begins to unfold: rootlet (radicle) down into the soil; shootlet (plumule) growing upward according to the laws of this other space, revealing *planar* organs, rather than radial or materially filled forms.

As well as being a little centre in Earth-space, the seed is the focus of an ethereal space, and gives birth, as the plant develops, to countless other foci of ethereal spaces, as the further buds and growing-points – seed-like organs – appear one after the other. They come into being, becoming visible in Earth-space, unfold and die away or become transformed as their purpose is achieved; and all in the manifold interplay of the cosmic rhythms with Mother Earth. It is an unending, rhythmic interplay between expansion and contraction considered *extensively* and contraction and expansion considered *intensively*. This deeper approach to the idea of "expansion and contraction" is very valuable; tiny forms, like seeds or pollen-grains, are ethereally vast, while large and fully developed forms are often nearing their end.

30 *Ethereal (Peripheral or Planar) Forces*

United with all developing life in organic Nature are the forces of the universe called by Rudolf Steiner, and also by an ancient and never quite obliterated traditional knowledge, the etheric or ethereal forces. These are the *peripheral* cosmic forces, a conception concerning which we can now form precisely, for with the ideas of modern geometry, we have transcended the one-sidedly centric approach to form, and are capable of becoming spiritually perceptive toward the signature of the peripheral forces and the planar space-formation of natural phenomena.

By means of an exact scientific method, we achieve an ideal basis upon which to build up an understanding of the polarity between the physical and the ethereal cosmic forces.

Negative-Euclidean space leads naturally to the concept of a type of dynamic forces, polar opposite or "dual" to those of classical physics and mechanics. These will be forces acting from plane to plane about the common line of the two planes, even as physical, point-centred forces act from point to point along the line which joins them.[34] This is consistent with the idea that the primary entities in such a space are planar. Since every plane in space – transformable as it is, projectively, as "vanishing plane" into or from the infinite – shares something of the quality of infinite expanse pertaining to the all-embracing plane, the planar type of force may also be described as *peripheral*, by contrast to the *centric* forces predominating in the material, inorganic world. This, then, is the essential theory we are advancing:

The processes of the spatial Universe involve not only the centric forces of which the prototype is gravity, but also the peripheral or planar type of force. "Negative spaces", interpenetrating the ordinary space of Euclid, provide the field of action for these peripheral forces, even as the latter space – its parallel and orthogonal structure determining the composition and resolution of material movements and physical forces — is the domain of gravitational, electromagnetic and other centric forces. The two kinds of space and force constitute a true, qualitative polarity — the primary polarity of the spatial world, more fundamental probably than the point-to-point polarities of physics. To this polarity the projective Principle of Duality (Polarity)[17] provides the ideal key. The phenomena we see around us are an expression of the interplay of the two opposite kinds of activity. In tendency, however, matter that falls out of the living process is predominantly subject to the centric type of forces, whereas in a living body, notably in regions of germination and vital growth, the peripheral type of force will be in evidence, in addition to and to a greater or lesser degree transcending the other.

Just as the space of Euclid and the corresponding forces are naturally described as physical space and physical forces, so too the negative or polar counterpart of these deserves a more concrete and descriptive name. We choose the name *ethereal*, thereby restoring to this word a meaning more akin to what it undoubtedly conveyed to our forebears — a meaning which still echoes on in literature and poetic diction. In this sense we speak of "ethereal spaces", meaning in the first place negative-Euclidean spaces as above defined p. 84, and of "ethereal forces", meaning the peripheral or planar forces. We here approach a wider question which is increasingly felt at the present time. Science as represented by its best exponents has grown more historically conscious, less categorical in its claims, more interested in its own cultural antecedents in pre-scientific times, and above all more alive to the problem. How does our scientific thinking, with all its technical results, affect the social and psychological stability of human life and even the cosmic balance of the Earth-planet as a whole? Now the idea of ethereal forces and activities belongs to the spiritual and philosophic heritage of mankind, both in the East and in the West. Needless to say, the ancient cosmologies did not express it in the scientific form which must be sought for if the conception is to be useful in the present age. But if one follows up in thought the geometrical paths here suggested — negative-Euclidean space and the idea of forces polar to those of classical mechanics — one enters in imagination a realm of spaciousness and buoyancy which is "ethereal" in the original connotation of the word.

Yet it will not only be a question of the re-discovery of something known in more instinctive ways to men of olden times. The opening of scientific knowledge in this new direction must also be a real spiritual achievement in the cultural development of modern man, and this will also call for a change of method, not to say a change of heart. In the science of living things, the separation between man the knower and the object of his knowledge is not and cannot be so great as in the science of the inorganic world. For man himself is born of living Nature. In mode if not in content, his thinking and imagination are a function of his own life as one of Nature's children. If the ethereal formative forces, of which the theory is here propounded, are a reality, they will be there not only in the growing plant or animal, the outer object of our researches, but in our own thinking activity, inasmuch as our own forces of life and growth have gifted us with power of imagination. If then our knowledge of the organic world, transcending the merely empirical and descriptive stage, is to penetrate to the *idea*, the underlying force and essential "law" of what is living, we are in a different situation than when examining the laws and forces of the inorganic world. We are a stage nearer to the primal fount, not only of the outer world but of our own thinking about this world. In saying this, the authors once again acknowledge their indebtedness to Rudolf Steiner,[13] who was the first to indicate that the "ethereal" be conceived and investigated in the present sense, and who outlined the methods to be followed. For he not only understood the essence of Goethe's method of research into the living world, but carried it a stage further.

Out of his spiritual researches, Rudolf Steiner described the ethereal formative forces, which sustain all life, as proceeding from the periphery inward towards a "relative central point", rather than from a centre outward. He described them as forces "which have no centre, but a circumference", meaning by circumference not just a horizon, but a whole sphere, – like the cosmic sphere of the heavens. In one lecture, he even used the words "surface-like" and "planar" to describe the forces working inward from the universe, and he described the "Gegenraum" as a "plastically formed space". Insisting that it is not possible to learn to know the "etheric or formative-forces body, which streams through the human being" by studying it from the point of view of ordinary space, he said: "It is only possible to study it, if we think of it as being formed from out of the whole cosmos; if we can understand that these *planes of forces*, approaching the Earth from all sides, come towards man and plastically mould his formative-forces body from outside."[35]

The "relative" central point, towards which the etheric forces "work from without inward" is the all-relating point or focus of the counterspace in question. It is *not the source* of the ethereal forces, but rather the innermost infinitude of the negative space – the infinitely receptive realm – such as is the outermost periphery of space for the outward radiating, centric, physical forces. The source of etheric forces is never a point-like centric realm, but a peripheral one; ideally it is a planar realm. What for positive space is most spread out and scattered, is for this other space the unified planar source from which the life-forces spring.

Ether-spaces are formed and dissolve again in the life-cycles of organisms. Wherever from a germ-cell – a seed, a germinating realm in an already created organism – new life unfolds, whether such a germ is in watery living substance or freely poised above it, we may discern the presence of an "all-relating point" – the inward infinitude of an ethereal space. This becomes evident where surfaces envelope and enclose the germinating point within, or through the gesture of leaf-like organs, which develop around an innermost point and open out from it.

The most elementary kind of ethereal force – a force of mutual *attraction* from plane to plane, in polar analogy to the gravitational attraction of material particles for one-another – will naturally be described as "negative gravity" or *levity*. This term again is justified by the expansive forms and movements which arise if one imagines what will happen, say, to a sphere enveloped and permeated by planar entities between which such a force is working. To each planar entity, in such a case, a certain intensity must be attributed – analogous to the *mass* of a material particle. We call it *levitational intensity*.[36] According to their intensities and geometrical distribution in the ethereal space to which they belong, a number of planar entities will have a resultant *plane of levity*, analogous to the centre of gravity of a material system.[37]

The hypothesis, that negative-Euclidean spaces and planar forces play a real part in living Nature, implies that the vast reaches of the spatial cosmos will have quite another function in this respect than in the merely pointwise, mechanical aspect of the world. For the infinitude is now within and not without, and as a corollary the origin, the middle region or focus of activity (it will be a *planar*, not a pointwise focus) will often tend to be in realms which from a physical point of view appear as inaccessible infinitudes. We will deal first with the geometrical and then with the cosmological aspect of this problem.

Geometrically, the study of an ethereal distribution will involve problems which do not occur in ordinary geometry or mechanics. For one is not pursuing the mere idea of negative space in the abstract, but observing how it is revealed in living forms, made manifest in physical effects. It will therefore in every instance be not only a question of the ethereal (negative-Euclidean) geometry as such, but of the way it is immersed in the physically spatial world. Often this will suggest the kind of form known as a plastic perspective – frequently recognizable in the plant kingdom – though with a dynamic significance which we are not accustomed to associate with a perspective. The simplest possible correlation will, however, lead to concentric forms, and precisely here the most important plane of the ethereal space will tend to be, physically speaking, infinitely far away.

For example, take a sphere or spheroid.[38] Projectively, the Euclidean centre is defined as the pole of the unique, infinitely distant plane. In ethereal space it is the *point*-at-infinity which is unique; a sphere will therefore have, not a central point but a central plane, polar to the unique point of the space. Since the expression "centre" implies a point, we have decided to speak of the mid-plane or *median plane* of the sphere. Precisely this plane will often be in the vast distances of the physically spatial cosmos. In effect, if the point-at-infinity is at the Euclidean centre of the sphere, the ethereal median plane will be the Euclidean plane-at-infinity itself – the uttermost periphery. In the simplest of living organisms – the spherical and polyhedral forms, for example among the protozoa – this concentric situation is clearly realized. Indeed it is most probably the primary situation, from which the more eccentric forms are evolved. The vast distances of the universe are therefore playing quite another part, in relation to life on Earth than would seem possible if the real structure of the world were only pointwise.

The same will apply in the realm of forces. Mid-point and median plane are purely geometrical concepts. Now in the physical-material realm, if a distribution of point-centred masses is symmetrical and uniform, the geometrical centre will also be the centre of gravity. For instance, in Plate X, top left-hand picture, if equal masses are placed at the ends of the radii on any one of the circles or of the spherical surfaces they represent, the centre of gravity of the system will be at the centre. The system of particles will tend to contract towards this common centre, and the picture might well be taken as representing successive stages of this gravitational contraction. Analogously, in the top right-hand picture, if the common centre is the point-at-infinity of an ethereal space and the tangent lines signify planar entities belonging to that space, we need but imagine them to be all of equal levitational intensity, and by their mutual attraction they will tend outward to their common "plane of levity", which will now be the infinitely distant plane of the physically spatial world. (Nor is there any need to suppose that they would take an infinite time in getting there, since the *ethereal* distance[39] which separates them from thence is by no means infinite!) This indeed represents the simplest, in a way the archetypal picture of a positively buoyant and expansive, levitating field of force.

Ethereal concentric spheres will, however, not always appear in this simplest form. If in relation to physical space the point-at-infinity is eccentric, they will appear as in Plate IX which shows in cross-section a family of spheroids with, say, U as common focus and with a common polar plane of U, appearing in the picture as directrix. In effect, they must all touch the absolute imaginary cone in U.[25] If this is spherical as we assume, a sphere in the ethereal space can only appear as a Euclidean sphere if U is at the Euclidean centre as in Plate X. The horizontal plane in Plate IX (considered spatially) is of course the common median plane; likewise it is the plane of levity if we imagine a uniform distribution of levitational intensities among the tangent planes of one or more of the spheres — uniform, that is to say, with respect to the measures proper to the space in question. This picture of ethereal concentric spheres with horizontal median plane beneath the point-at-infinity is archetypally related to the gesture of unfolding leaves at the growing-point.

The idea which is shown in Plate X (top right), with its particular relationship between "centre" and periphery, has, however, archetypal and cosmic significance. We will state it in the following words:

An ethereal-concentric formation, with a common median plane, appears in physical space as concentric, when and only when the common median plane is the infinitely distant, all-embracing plane of physical space. Then, however, the central point of the physical-spatial formation is at the same time the all-relating point of the ethereal formation. A form is concentric both in space and in counterspace when and only when what for the one space is the all-embracing is for the other space the all-relating Absolute.

What is here formulated in pure thought corresponds to real relationships in

Nature. All forms which have arisen out of a living process, and become visible and tangible in a physical body, have arrived, as it were, in physical space, and are things among things. They are to be found in a particular place on Earth and with a particular relationship to the Earth's surface.

If such a created form has come into being out of a formative ether-space, then, in the gesture and quality of its form can be recognized its relation to this ether-space. But there belongs to this thought the question: What relationship to physical space has the ethereal median plane belonging to a particular form?

If the organic form reveals a concentric symmetry, as do many of the minute protozoa, studied so intensively and pictured by Ernst Haekel (Figs. 56 and 57), it is in a special sense cosmic. These forms, with their wonderfully regular, polyhedral shapes sometimes have concentric shells, one within the other. The median plane of the ethereal space-formation is here at the same time the heavenly sphere itself – the all-embracing plane of earth-space.

This comes to expression mostly in organisms which float freely in water or which develop in a fluid medium, such as the germ-cell or morula in the early stages of development of a complicated organism.

In this more concentric growth, Nature reveals the meaning of what we have called growth-measure, namely, the result of a *balanced* interplay between an inner infinitude and an outer, cosmic periphery. The seed or germ-cell is an "infinity within". Here we touch on the morphological secret which divides the inorganic from the organic. Its signature is revealed in the "Spiral of Life".

Many organisms reveal this physical-ethereal spatial concentricity, not so much in all three dimensions, but in a plane, or projected into a plane, for instance, those with axial symmetry. The plane in question will then lie at right-angles to the axis of symmetry, and its infinitely distant line is polar to this axis or – to an "all-relating point" within the axis. These, for example, are "ethereal median-planes", in which the symmetry may be revealed. The cyclic or regular spiral-formations of many leaf-rosettes and in the centre of many flowers is of this nature. It is also revealed in most sea-shells (see Figs. 52–55).

In the gesture of living forms, we learn to read the script of organic morphology and see the fundamental difference between those forms which lie inert in Earth-space and those into which there plays – or has played – the cosmic force of negative-gravity or "levity", drawing the living substances upward towards the light and air and away from the earthly centres of gravity.

32 Cosmological Aspect

Returning now to the simplest picture, where the ethereal forms are not only concentric in their own right, so to speak, but are concentrically planted in the physical universe of space, we have to face the cosmic implications. At first sight the idea of a realm of form and of dynamic forces having their source or planar focus in the infinite periphery seems paradoxical. For one instinctively tries to relate it to the existing physical cosmology with its vast distances – parsecs and

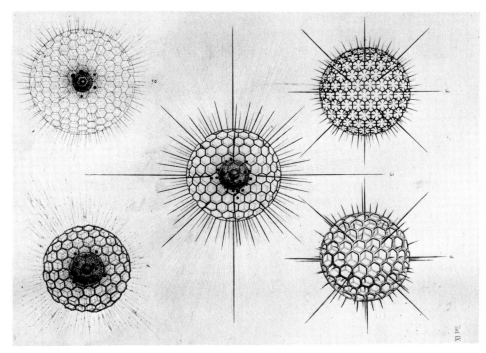

56   Radiolaria (from Haeckel)

57   Radiolaria skeletons. Sphere and Radius

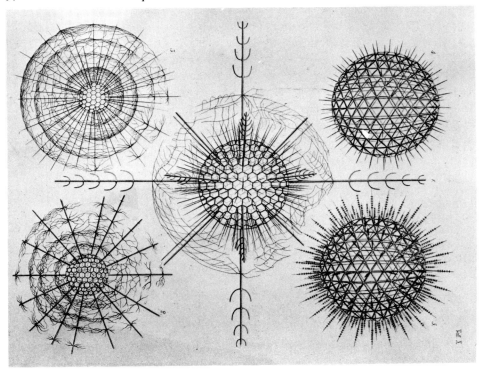

light-years – estimated, in the last resort, in terms of miles or kilometres. It is of course a real question, what the relation of the ethereal formative spaces to the universe of stars will be. But it is probable, as has so often happened in the history of science, that we first have to learn how to put the question. We must go one step at a time. It is in the living world on Earth – the plant world above all – that we see those phenomena which find their natural interpretation in the idea of negative or planar space. There is no reason to abandon this because we do not yet see how it relates to the existing physical cosmology – subject as the latter is today, in any case, to rapid and far-reaching changes – or for that matter to the existing theory of the sub-microscopic realm of molecules and atoms.

We are, however, attributing a fundamental role to the Principle of Polarity in the real nexus of the universe, and this in two respects can help us with the present problem. On the one hand the two polar-opposite aspects of truth which it reveals can each be studied in its own domain. Undoubtedly the interrelation of the two, when understood, raises one's knowledge of the truth on to a higher level. But it does not invalidate or lessen what was known before, of the one or the other aspect. Over two thousand years of "Euclid" taught men the truths of pointwise space; these were in no way vitiated, only their understanding was deepened, when the polar aspect was revealed. To quote Professor Turnbull,[40] the two aspects are not against, but "in and through" each other – "contemporaneous and complementary . . . Nevertheless, each can be followed for its own sake without necessarily forcing attention on the other." Therefore we may allow the phenomena of life to teach us concerning Nature's planewise aspect without embarrassment due to the vast accumulation of "pointwise" knowledge, and above all, pointwise theoretical constructions.

On the other hand the Principle of Polarity shows that the different entities of space are not external to one-another but interwoven. For point- and planewise space, not only are the standards of large and small reciprocal to one-another; even the relationships of part and whole are interlaced. In Euclid's pointwise aspect, for example, the point is the elementary entity "which hath no parts"; the plane is composite, a two-dimensional manifold of points. In negative space the roles are interchanged; the plane is now the fundamental element and the point is composite – a two-dimensional manifold of planes. And this is only the beginning; taking the lines into account, the mutual relations as to part and whole are even more deeply interwoven.

Projective geometry thus teaches an organic world-conception, transcending the merely additive idea of size and structure and the crude eighteenth-century idea of spatial causation, where outward and impenetrable objects – real or imagined – do but push and pull each other. But the same lesson is to be learned from the phenomena of life, if we can look as Goethe did with fresh and open mind. Not only gravitation and inertia, by the Newtonian interpretation (or by its present-day modifications) ruling the cyclic movements, show the Earth-planet's community with Sun and stars. The life of plants reveals it no less directly. Through the green plant, all life on Earth is sustained by the Sun's light – a gift from the universe of stars. Nor need we merely think of the solar energy as outward spatial causation, striking the green leaf after having made its way through the intervening mileage. What the phenomena reveal and what healthy feeling tells as we accompany the Earth's life through the seasons, is reinforced by the ideal lessons of the new geometry. The plants will now appear rather as organs which the Earth puts forth, expressing in a primary and direct way her organic relation to the Sun and through the Sun to the celestial universe. When we begin to look at the question in this light, the finding of planes of levity in the vast periphery of the cosmos no longer appears so paradoxical.

The concept of ethereal space enables the Earth-planet as a whole to be regarded as a living entity – not in a vaguely philosophic sense, but in a way that lends itself to detailed investigation. As a first working hypothesis, we imagine the ethereal space of the planet to have its point-at-infinity at the Earth's centre. The tangent planes at the surface – in the concentric spheres at different levels (stratosphere, atmosphere, hydrosphere, biosphere, lithosphere, etc.) – are then no mere geometrical abstractions but represent, potentially at least, constituent elements of the ethereal planet, just as the stones and grains of sand, each with its centre of gravity, are among the constituents of the material Earth. The mutual force of attraction between these planar entities will then constitute a field of levity, in polar relation to that of terrestrial gravity. Physical entities will under certain conditions be received into the sphere of action of these ethereal forces, resulting in phenomena which are not due to physical or pointwise (atomistic) forces only. A primary phenomenon of this type, according to the theory here advanced, is the upward and outward growth of plants (compare Chapter V).

If the distribution of levitational intensities is uniform, the Earth's plane of levity will be the plane-at-infinity of physical space, just as the point-at-infinity of the ethereal space is at the same time the centre of gravity.

The question is at once suggested whether this simplest of relations – namely the mutual concentricity of the Earth's gravitational and levitational fields, or of the physical and ethereal spaces to which the planet belongs – is subject to geographical modification and, above all, seasonal variation. The phenomena of plant life suggest a seasonal swing in levitational intensity between the northern and southern hemispheres. The plane of levity itself will then be undergoing some regular form of cosmic movement, for which the plane-at-infinity may represent not the stationary but the average or equilibrium position.[37]

Regarding the planet as a whole in this ethereal or planar light, phenomena which appear merely physical when studied in minute pointwise regions and with pointwise forms of thought, as for example within the confines of a laboratory, may reveal quite another aspect when the entire planet or even wide geographical areas are considered. We believe that this will prove so above all in the hydro-, aero- and thermodynamic realm, so that the concept of ethereal space and force will open up new prospects in meteorology.

If an ethereal field is to be attributed to the Earth itself, the same will apply *a fortiori* to the Sun, source of all life on this planet. (We leave aside, for the purpose of the present work, the question of the other planets, also the relation of the

Earth's ethereal field to lunar rhythms.) Here the well-established physical theories – theories of light and radiant energy generally – will at first occasion difficulties. From the phenomena it is quite evident that the ethereal will have to do above all with light and warmth and the chemically active rays, which play so great a part in physiology and of which, for life on Earth, the Sun is the primal source.

33 Organic Interplay of Spaces

It becomes a methodic principle to look at all things from the polar aspect. Take once again the simplest instance, suppose the form to be spherical, with or without a sharply defined outer surface. Imagine the top left- and right-hand pictures of Plate X superimposed concentrically on one-another. This then would represent a spherical entity at once ethereal and physical in nature and indicates the form of question. To what extent does the one or the other predominate? Or is it – as a limiting case, so to speak – wholly belonging to the one realm?

If the sphere has a well-defined surface and therefore size, we shall not judge this from the mere physical aspect as though it might grow bigger and bigger into the surrounding void; we shall experience a certain balance between outward and inward, radial and peripheral magnitudes. In its ethereal aspect the sphere grows bigger when the surface recedes towards the infinite point within. The physical and ethereal aspects of magnitude are reciprocal. This applies, in particular, to the normal size of any living creature. Growth is an interplay between two infinitudes, both of which express degeneration or loss of form – the one by contraction into a point, the other by flattening into a plane. (This is most vividly portrayed in the picture of "ethereal concentric spheres", Plate IX.) Each living entity according to its kind achieves "generous proportions" when in its own way poised between the two extremes. It is a balance between *two* active principles, physical and ethereal respectively.

The same polar aspect arises – at least as a question to be put to Nature – with regard to the manifestations of radiant energy. Think of a radiating centre – a glowing flame for instance, or a space in which some chemical reaction is generating heat. May this phenomenon too be a manifestation of ethereal space as well as physical? If it be so, the surrounding periphery of space is no mere void into which physical energies are being spent, but is playing a more active part.

In its planar character and in its orientation towards an infinitude within, the concept of negative-Euclidean space provides, as was said, an essential key to the formative processes of life. This does not mean, however, that it can be applied as rigidly as can the laws of ordinary space and of the physical forces working upon inorganic matter in solving, say, the problems of mechanics. The living realm is more transient and mobile. In all but the simplest of living organisms there will be more than one ethereal centre or point-at-infinity within; in other words there will be a number of ethereal spaces interpenetrating one-another, according to the differentiation of organs and functions. Their formative processes will be related organically and in a definite hierarchy of importance. Some will be subservient to others; some will be relatively transient, others more enduring.

Geometrically and dynamically, this involves problems; but our guiding thought is in the right direction. The underlying notion is that of projective space. This becomes differentiated according to the spatial character of the "absolute" which is presumed invariant. If the "absolute" is composite, or even subject to variation in time, the problems that arise will once again relate to the more general projective space. In thinking out the negative or "dual" of the space of Euclid, and in conceiving that this type of space is essential to the formative processes of life, it is therefore not a question of attributing to the latter all the rigidity and permanence of the former. One may let go of a fixed anchorage while still retaining the fundamental concepts of projective geometry – the interrelation of point, line and plane. This science as developed hitherto – notably in the theory of continuous transformation-groups, where the invariant elements are of very diverse kinds – is rich in morphological possibilities, clearly akin to the phenomena of living growth.[41] But one must first be freed from the one-sidedly pointwise bias which is as foreign to this more mobile geometry as indeed it is to the phenomena themselves.

Science asks the question: how by the convex and outward growth of each single cell, by cell-division, does it come about that the microscopic parts grow in precisely the extent and the directions which the organism as a whole requires? More and more as time goes on, the scientist begins to think in terms of fields and areas of biological activity in mutual interplay. But also an important reversal of spatial imagination is here required.

Where the idea of physical space is instinctively, exclusively maintained it goes without saying that the larger whole will not be understood until the microscopic parts are; they, after all, are like the bricks of which the whole must be built up. But this is not the logic of ethereal construction: here on the contrary we are more likely to learn the elementary truths by contemplating the entire form, visible to the naked eye, and, from the lessons thus learned, in course of time press forward to understand Nature's technique in making the microscopic parts serve the whole. We may conceive that there are a multitude of "differential" ether-spaces following an archetypal pattern – each with its "infinitude within", – their phases of activity waxing and waning to serve the whole.

In every living cell, in every embryonic tissue, there is a primitive potentiality of growth. Where the true balance of a living body is impaired, often a local tendency to spherical and outward growth will become rampant. In the healthy organism this tendency would be held in check, only allowed to function in so far as it served the whole. We may thus gain a new conception of a well-known biological fact: the greater or lesser predisposition of a higher organism to infection or parasitic attack by a lower.

The archetypal germinating centre is none other than the point-at-infinity of an ethereal space pure and simple. To live at all, in every growing part the higher organism must be giving birth to such centres. If it is healthy they are held in balance, inducing their several outward growths only to the extent that the organism as a whole requires. But if in one part or another the localized ethereal centres become too strong, it may only be a relative question whether the excessive life finds vent as it were of its own accord, or whether the ethereal focus becomes occupied by the specific seed or spore of some other organism, bearing another morphological archetype and able to make use of the unbalanced ether-forces of the host.

The geometrical idea of spatial forces – pointwise or planar – and of their static and dynamic balance does not lose all definition at this level. Even in the mechanical realm, for instance in the statics of a rigid body, there are important truths of a purely projective nature – invariant not only in the rigid space to which these particular forces belong, but under all projective transformations.[42] This shows that the wonderful harmony between the forces of Nature and the forms of space is more deeply rooted, and it encourages the further development of a purely projective theory of forces – a subject to which mathematicians have already devoted some attention.

Moreover, it is precisely at this level that science will one day understand the relationships of the microscopic with the astronomical phenomena. The question of the relation of the Earth's ethereal field to lunar rhythms, to the other planets, as well as to the Sun, will find answers, when science has conceived the true Idea of an organism: a composite field of more or less balanced interplay between *polar* formative forces and principles. In this book, we shall concentrate to begin with on the fundamental polarity of Earth and Sun, for this is archetypal. When this idea has been fully experienced and scientifically understood, the way will be open to perceive in detail how the true nature of the plant reveals also the true nature of the galaxy – of the great organism within which all life on Earth takes place. The plant must reveal herself in all her glory – an ethereal form clothed in earthly substance, living, unfolding, metamorphosing between "Earth and Sun", amid the *sun-like*, though differentiated cosmic forces of the planets and the stars.[43]

34   Cone as Intensive Form in Counterspace

The ethereal counterspace is, like physical space, to be thought of as being three-dimensional – only negatively; it is to be thought of planewise from without inward, instead of pointwise from within outward. In physical, Euclidean space, the two-dimensional element is the plane and the one-dimensional the straight line; it is natural for the understanding of this space that one learns two-dimensional geometry in a Euclidean plane.

In three-dimensional counterspace, there are two- and one-dimensional aspects also, but here the extensive and intensive spatial experience is reversed.

The one-dimensional is still the straight line, but as a manifold of planes instead of points (Fig. 28). The negative two-dimensional is the *point* as manifold of lines and planes. Here we have an *intensive* two-dimensional element, polar to the extensive, two-dimensional plane of Euclidean space. Thus there are not only polyhedral, discrete forms (as, for instance, in Fig. 26), but also plastic forms resulting (as in Fig. 42) from a continuous sequence of lines (so-called generators) and tangent planes or planes of contact. These are *conical* forms of various kinds, held in a point, which are polar to the plane-curves of ordinary geometry (see also Plate IV).[20]

Take, for example, the spiral curve drawn in perspective at the top of Plate VI, which is like the plane-spiral in Fig. 51. Out of it there arises the spiral cone in the point below. (The spiral above is only drawn to make visible this cone formation, at least to some extent.) The conical surface spirals in infinitely many furls towards the innermost – in this case, vertical – line; just as a plane-spiral uncurls towards the infinitely distant line of its plane. Following the conical surface as it unfurls (impossible to include in the drawing), it opens out, becoming more and more horizontal, towards the horizontal plane indicated in the drawing; just as a spiral in a plane curves ever inward towards its mid-point.

The spiral cone in a point curves between the extremes of the inner-most line and the plane as outer-most element.	The spiral curve in a plane curves between the extremes of the outer-most line in the periphery and the innermost point.

The concentric circles (Plate X top left) have their polar reciprocal picture: a family of concentric circular cones (Plate IV). As the circles grow infinitely outward in growth-measure towards the infinitely distant line of their plane, or inward towards the centre, so the cones grow "inward" towards the vertical middle line and "outward" towards the horizontal plane.

In Euclidean space, there are two kinds of *plane*: the infinitely distant one and all the others. Their measures and form-relationships arise from the fact that in each plane a single line governs, namely, the line that plane has in common with the infinitely distant plane. (In the infinitely distant plane itself, *all* lines are infinitely distant. Through this, and because of the imaginary circle, there are other measures related more with spherical trigonometry.)

In counterspace, there are two kinds of *point*: the all-relating point and all other points. Also here, every "other point" bears a *singular* line, namely, the line which that point has in common with the all-relating point. This line also *functions* as an infinitude, but from within rather than from without. We experience it clearly as an *infinitude within*, and precisely this concept is a significant guide towards the understanding of the ethereal nature of the forms which appear in the growing shoot of the plant.

The cone forms in Plate IV are to be regarded in this sense; they are forms in counterspace. The point bearing the cones is *not* the all-relating point; the latter would be somewhere on the vertical line *above* this point. (How far above is for the moment immaterial.) This line, then, represents the *line-at-infinity within* of the "cone-space". For the concentric circular cones in Plate IV, the horizontal plane towards which the cones open out, is the "ethereal median plane", and is comparable with the central *point* of a family of concentric circles in physical space.

In this sense, the family of concentric cones is to be thought of in counterspace. Following them as they become ever slimmer and slimmer, from without inward, we should *not* experience them as shrinking; on the contrary, it is an inward growth-process, towards the line-at-infinity within, exactly comparable to the growth of the circles in Plate X (top left) outward towards the infinitely distant line. Change the direction, and the circles shrink into their mid-point; this compares with the opening of the cones into their median plane. This opening out in the counterspace must be experienced, not as outward *growth*, but as a *lessening* in the cone-space; as such it disappears into nothing, as it flattens completely into the *plane*, just as the circular space in the plane of the concentric circles is reduced to nothing in the central point. (Compare paragraphs 20 and 21, Chapter III.)

We are so easily confined to words which express the familiar properties of a physical, pointwise space. As well as the word *concentric*, we must become used to the word *co-planar*, which indicates, not a central point, but a planar, *peripheral* middle realm of an ethereal space.

The ethereal, negative two-dimensional geometry in the point is of infinite significance in the study of plant morphology and growth-processes. When Goethe writes to Herder: "Forward and backward, the plant is ever leaf, inseparably united with the germ of the future plant, so that the one is unthinkable without the other", the relating of *leaf and germ*, or *leaf and eye*, is ideally none other than the relation of the *extensive and intensive two-dimensional elements*.

In relation to the Goethean conception of "expansion and contraction", we quote here the Dutch botanist F. H. Julius,[44] who wrote: "The outer, visible rhythm of growth is interpenetrated by another, invisible rhythm of growth-*forces*, which runs in just the opposite direction. . . . The more contracted the material, the more intensive are the possibilities of growth. The more the material expands, the more these possibilities ebb away." Julius calls the realm of the creative, formative forces *"the realm of the Idea"*, which lies at the fount of the becoming and growing of the living organism. In this sense he goes on: "There is also to be found expansion and contraction in the realm of the Idea, but not spatially; rather as a greater or lesser potential of creative force. Where there is spatial expansion, there is ideal contraction; where spatial contraction, there is ideal expansion." With "spatial" is meant here, *positive* space, expressed in the usual terms. If one is familiar with the relationship of the formative forces with ethereal or negative spaces, then in these words exactly that is described which is revealed to perceptive judgement, when one lifts the experience of the spaces and forms of the plant-shoot to that realm in which one also perceives the ethereal spaces as such.

To understand the plant according to this way of thinking, we must become more familiar with a sun-like space. Until now we have concerned ourselves mainly with the spatial-counterspatial *form* as such, and have hardly touched on the aspect of substance. In positive space, the volume, say, of a sphere or a cube provides a natural vessel in which to contain substance. It is, however, in the nature of the case that such finite, spatial contents do not accumulate in a counterspace determined by an all-relating point.

The question arises: Is there for the ethereal spaces something analogous to the filling of physical space with material substance? Spiritual science answers this question in the affirmative. Let us answer it, to begin with, with a scientific *hypothesis*, which, if we are content to follow the direction of the idea, can lead to some enlightenment in the contemplative experience of natural phenomena. We will formulate the idea somewhat radically:

The filling of physical spaces with ponderable matter represents only one pole in a polar process. It cannot be that in some places in the universe ponderable, positive space-filling matter accumulates, if not somewhere else a balance is being held. In the totality of universal processes, there is physical, weighable matter on the one hand and "negative", imponderable, – anti-ponderable – "ether-substance" on the other.

Negative substantiality is in its spatial aspect planar; it "fills" space planewise. Using the two upper pictures in Plate X as a guide, think of a physical sphere filled to its surface with ordinary matter. Then picture the polar opposite sphere; here the "negative substance" formed of planes "fills" the space from the world-periphery inward, as far inward as the surface of the sphere. What would normally be called the inner space of the sphere remains empty; it is hollow. The space filled with planes from the world-circumference inward is, *considered as negative space*, finite; it is *not* infinite in content. Infinite is, however, the hollow, inner space of the sphere, for there in the heart of it is the all-relating star-point, the infinitude within.

Leaving open the question as to whether quantitative concepts have any meaning at all in relation to ethereal substance, we can at least be clear that according to the purely mathematical idea, the planewise space of the sphere, from the periphery inward, is certainly finite.

The same concept comes to our aid, when we think of a materially filled sphere, like the Earth, and then consider how the material aspect becomes less and less as we pass out through the Earth's atmosphere towards the world-periphery. Correspondingly, then, one can form the opposite picture: a sphere with a maximum of intensity in the periphery, which tends to diminish and gradually disappear towards the all-relating point in the infinitude within.

One must recognize that somewhere in a living organism there may be a point or a localized region, functioning as focus for ethereal-substantial content. Then the laws described in paragraph 30 come into question. In such a realm, we do *not* look for the ethereal *content*, but rather for a *receptive* realm, a space infinitely capable of receiving the ethereal content, which according to its nature, fills the surrounding, peripheral space.

For the concept "substantial content" in physical space, we have to substitute the idea of a *nothing — nothingness* — in ethereal space. It is an empty, hollowed-out space. There, where in physical space one is used to finding substance, there is *none*; whereas in the surrounding space, where one expects to find an empty nothing, there it *is*. The sun-space, or counterspace is in relation to ordinary space a *negative space*. The two are related as the stamp is to the seal.

Fig. 58 shows some enlarged photographs of microscopic sections through buds in very early stages of development, showing the rudimentary beginning of growth in which even the flower-bud of a future season is being laid down; Cranberry (*Vaccinium vitis-idaea*), Cherry (*Prunus avium*) and Arctic Blackberry (*Rubus arcticus*). Such pictures speak eloquently to the imagination concerning the hidden, inner, receptive centres — the ethereal infinitudes deep in the womb of Nature.

Each time a man plants a seed in the Earth, it is not the little pointwise seed which contains the ethereal substance. In the seed lies the receptive centre, *thirsty for form*. And the spot on the Earth where the seed is planted is not the place where the ethereal substance is to be found, for this substance is not pointwise, it is planewise, peripheral substance. Here, where the plant will grow, become visible, and show forth its ethereal organs, is the hollowed-out, negative space, the inward emptiness, into which the ethereal forces will stream, to fill it with living, nutritional substance. We touch here the realm of divine *magic*, and may understand that in less materialistic times than ours, the deeds of sowing and reaping are accompanied by acts of prayer and worship, in the recognition that here is a realm of human activity, which reaches beyond man into the spiritual depths of all earthly existence.

36   Weight and "Leichte" — Earth and Sun

The concept of interwoven polarities now enables us to look more deeply into the greatest polarity of all — that of Earth and Sun. To the Earth, physical and material as it obviously is, we have also attributed an ethereal space, with the point-at-infinity at or near the Earth's centre. We saw it in terms of the two pictures of Plate X, placed one upon the other concentrically, but with the left-hand picture predominating. In the same dual aspect we now see the Sun, only with the ethereal — the negative-space quality of the right-hand picture — predominating. The sympathy, the organic, life-begetting interplay of Sun and Earth is due to the fact that each in some degree shares the other's nature; none the less, they are polar to each other, so that a mighty activity is engendered. If Earth and Sun, as most cosmologies agree, were once united, the bonds were loosed in such a way that the Sun took with it a preponderance of Light, and the Earth of Darkness. But they were not irrevocably sundered, and through the countless seasons of Earth's separate existence, life on this planet still bears witness to their kinship.

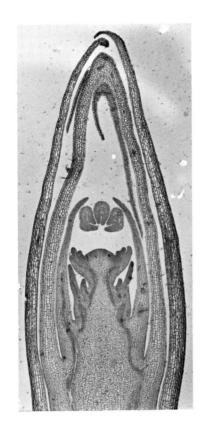
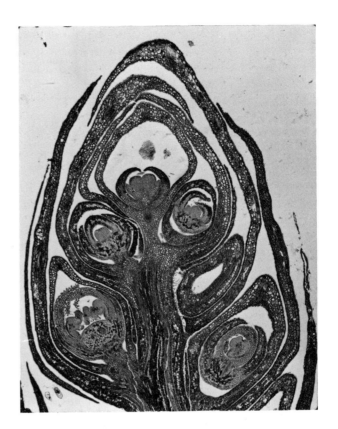

58   *Microscopic Sections of Developing Flower Buds*

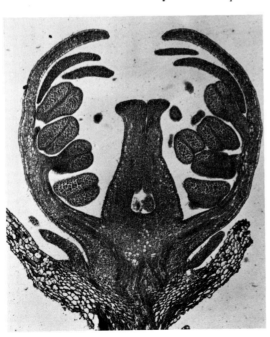
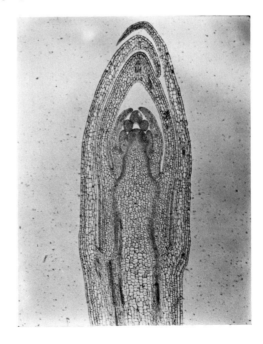

The concept of ethereal space, interpreting what is known of living form and function – and above all, of the plant in its relation to the sunlight – reveals a new and fundamental aspect of the sun itself. It is as follows:

In the spatial universe of which we have experience on Earth, ruled as it is by the great polarity of point and plane, the ethereal or negatively spatial aspect has its most essential, macrocosmic focus in the centre of the Sun. Mighty peripheral activities pour in towards this macrocosmic focus. Of all ethereal spaces, this of the Sun is the most powerful and the most ancient. It is organically related to the planet Earth, and to the living creatures which the Earth brings forth. Therefore there is a primary affinity between the ethereal spaces formed by living organisms upon Earth, and the vast cosmic space of the Sun itself. The life-giving power of the Sun's radiations is due, not only to outer physical causation making its way hither through external space, but to this primary, inherent kinship.

The proposition has been stated with all care. Therefore we say, at the very outset, "in the spatial universe of which we have experience on Earth". From the aspect of other stars and galaxies, although their light is manifest in our space, it may be different. Again it is not implied that the ethereal aspect of the Sun is the only one. The proposition contradicts no established facts, such for example as the relation of Sun and Earth in the aspect of physical space or of "celestial mechanics"; it adds a complementary aspect, leaving to future discovery the question of the full relationship between the two.

In the mechanics of the physical world, the most important concept is that of a "centre of gravity". For every material body or given distribution of matter, this is a definite, even calculable point, between which and the centre of the Earth we conceive the force of gravity to work. For the community of all earthly-material bodies, the centre of the Earth is like an archetypal centre of gravity to which all others relate.

Physical matter is ruled by the force of gravity, the archetypal instance of the centric forces – and by those forces closely allied to it, such as pressure and contraction. These forces express themselves ultimately in the power to move matter or to withstand and counterbalance the mechanical type of force, which is also characteristically centric.

We have been led to the clear conception that the primary force of an ethereal space will be the precise opposite of gravitation, or of a force of pressure. It will be levitational, suctional, expansive.

Once again it is difficult to find words to express adequately the unfamiliar concepts and phenomena, and, as so often, the more flexible German language provides the best answer. The German word "*Leichte*", which is not quite the same as "Leichtigkeit" or "lightness" has a quality, which makes it preferable to any other, and we like to use it. The ethereal forces are not light and empty, in the sense simply of a lack of physical weight; they have their own kind of "*anti-weight*", which the word "*Leichte*" comes nearest to expressing, – better, perhaps, than the word "levity".[36, 37]

Thus, analogous to the centre of gravity of a material body we can speak of a plane of Leichte (or levity), of a form of negative substantiality, in other words, of ethereal content. A median plane in the realm of ethereal substantiality holds the balance over against the concept of a mid-point in materially filled space. And as we have seen in paragraph 31, through the mobility of projective geometry, we are not restricted to the idea of ethereal concentric spheres, as depicted in Plate X, where the Leichte-plane is the plane at infinity of physical space itself. We recognize it again in the excentric form of ethereal concentric spheres as represented by the horizontal line (Plate IX), and also in the horizontal plane in the "Cone-space" in Plate IV.

It is helpful to consider in this context the concepts: *centrifugal* and *centripetal*. In regard to physical matter, gravity functions centripetally. Nevertheless, through the resulting tendency towards concentration of matter, the opposite quality is manifested; for all matter, as soon as one tries to lessen the space it occupies, reacts with outward pressure. Both qualities are inherent in all matter and this comes to expression in the "specific weight" and the "specific volume". In solid substance it is largely the former which predominates, in airy or gaseous substance certainly the latter, where, in the absence of a pressure from without, the space-filling tendency continues more or less indefinitely.

It is in the nature of the case that in the filling of space with matter, the one aspect, however dominating it may be, cannot be present without the other. For example, were the force of gravity alone to operate, every material body would collapse into its own centre of gravity – or towards the centre of gravity of the Earth. The centrifugal tendency must work in some way or other against the centripetal force. (According to the theories of modern physics, it is the hypothetical "intra-molecular force" which plays this part. We are, however, not concerned at the moment with the theories, but are simply considering the facts of the phenomena.)

In the case of ethereal substance in counterspace, we must think in the opposite way. Leichte works *centrifugally*, drawing or sucking living substance upward and away from the all-relating point or relative centre of the living process, towards the Leichte-plane. Here, too, there is the balancing, centripetal process, for the ethereal forces *seek* the relative centre, their tendency is *to give, to bestow* their life-awakening qualities. The ethereal forces do not simply follow their natural, cosmic tendency, and float away into the periphery, but offer their living forces inward, towards the living, growing centre, the infinitude within.

The peripheral, etheric forces work inward and unite with the physical, but in doing so, the levitational force draws or sucks the physical outward. The expansive process does not come about through pressure from within, but through a planewise "suction" from without. One could formulate it thus: *The physical, held by gravity, draws the ethereal in towards itself (germination, invisible prelude to growth). The ethereal, carried by Leichte or levity, draws the physical out with it (outwardly visible growth).*

We become aware that the concepts centrifugal and centripetal have a double meaning in this context; they correspond to the qualitative process and not merely to the outward appearance. The idea of expansion and contraction in the realm of growth is more subtle in the Goethean sense than in the Newtonian. The true polarity in the realm of the living is the expression of the centrifugal in the one realm, together with the centripetal in the other, and vice versa.

Physical	Ethereal
Centripetal: Gravity	Centrifugal: Leichte
Contractive	Expansive
Tendency to collapse into its own centre of gravity	Tendency to rise towards its own plane of Leichte
Centrifugal: pressing outward Dominating its own space from the centre outward	Centripetal: drawing inward Giving bounteously from the periphery inward
Physical Substance	Ethereal Substance
Convex (pointwise) filling of space Orientation outward towards the all-embracing plane, outward: the "empty space"	Concave (planewise) filling of space Orientation inward towards the all-relating point, inward: the "hollowed out" space

In the science of cosmology, it is of fundamental importance to allow the hypothesis that in the real processes of the universe there may be both positively and negatively filled spaces.[45] One meets the phenomena of the spatial cosmos with quite other expectations. Until now, the cosmological theories of the scientific era were based quite as a matter of course on the idea of a one-sidedly pointwise, Euclidean-type universe. In spite of the general Theory of Relativity and various other developments in recent time, this deeply rooted presupposition has not changed as radically as the conception of space and counterspace would require.

Concerning the significance for science of the concept of Polarity in the so-called "Principle of Duality" in modern geometry, the English mathematician Professor Turnbull[40] (who published the letters of Isaac Newton) wrote: "The two aspects are not in contradiction, but are mutually in one another. They are complementary and simultaneously true. And yet it is possible to research into either realm without finding it necessary to keep the other in mind at the same time." Professor Turnbull draws attention to the new morphological method employed in this book: "In the formation and growth of organisms, both modes of analysis are important (by this is meant the pointwise and the planewise method of observation). Seed, stem and leaf ask that the three-dimensional plant form be studied in two aspects, – not only pointwise and microscopically, but also planewise."

The ideas of the new geometry lead to an organic world conception. The external, additive concepts of size, and the rather primitive eighteenth-century concept of cause and effect, which still persists, in which entities in space push and pull one another, are overcome. The laws whereby the planet Earth and the Sun are related are described not *only* according to the spatial astronomical thought forms of Kepler, or according to the laws of inertia in the sense of Newton or even Einstein. They are also revealed by the living plant, when we learn to understand her laws aright.

The ethereal aspect of radiant energy will be revealed in its life-giving functions. In the organic cosmology resulting from the fundamental concept of spaces positive and negative, the relation of the Sun to the green plant on Earth will guide us in discerning the ethereal character of the Sun itself. What the polarity of space suggests will here go hand in hand with a very simple observation of phenomena.

In physical space, the apparent source of radiation is, to begin with, a point-centred body – a candle-flame, a glowing ember. With its potential effects the radiation fills the surrounding space, tending towards the infinite periphery, the effects becoming less intense in proportion to the inverse square of the Euclidean distance from the source. The radiation becomes perceptible by its effects on matter. It is reflected from the surface of material objects. Absorbed in these or in the medium that fills the space, it gives rise to heat, raising the temperature and thus inducing other changes. Its relation to dark matter is symptomatically shown by the Bunsen flame. Non-luminous and scarcely visible until the stream of air is cut off, the flame then grows relatively smoky and at the same time shines forth brightly, the dark particles of carbon becoming incandescent.

In physical space and in relation to matter, radiation manifests centrifugally. This is a simple description of the phenomena, quite apart from any ascertainment of the so-called "velocity of light". It fills the space to the remotest material barrier. The heat which it engenders when absorbed brings about thermal expansion and nearly always increases solubility; in the solution it enhances the outward, space-filling tendency known as osmotic pressure.

If we now try to imagine the ethereal or polar counterpart of this centrifugal picture, it will be a radiant activity of which the active source is in the infinite periphery of space. Once more we have the two contrasting pictures of concentric spheres in Plate X, of which the left-hand one clearly represents the naïve physical idea of radiation. On the ethereal side we have to picture the concentric spheres not in their pointwise or radial but in their planar or tangential aspect. The planar tendency is inward and centripetal, but in describing it as such there is no need to think of it realistically as though the planar entities were *travelling* inward. The relation of space and time in this realm is on another level.[46] The essence of the matter is that the active source is in the periphery, whilst in the centre, where in the aspect of physical space we should expect the source, is the infinitude into which the ethereal or negative radiation is being spent.

Turning now to the phenomena, it is precisely this centripetal tendency which the processes of life, in their relation to radiant energy, suggest. Apart from what comes to Earth directly from Sun and stars, the primary source of radiant energy as known to men on Earth is fire, burning the fuel – wood or coal or oil – which has been formed by living things. Directly or indirectly, it is the outcome of the relation of the green plant to the cosmic sunlight. Now the formation of sugar, starch and cellulose in the leaves and other organs of the plant is a phenomenon, not of expansion but of contraction. The chemical constituents of which these carbohydrates are formed – carbon dioxide of the air and water raised from the soil – are specifically lighter than the resulting product in the body of the plant. The chemical process itself is endothermic and therefore cooling and contracting. Even the polymerization, forming the di- and polysaccharides, the different forms of starch, cellulose and lignin, is a condensing process.

When afterwards the wood is burned, radiant heat and light are emitted, and in its physical manifestation the outcome is once more expansive. Thus the relation of the cosmic sunlight to the green shoot suggests that *in this living process light is associated with a centripetal activity*, to which the building of these life-sustaining and combustible substances is due. The release of radiant energy with its expansive physical effects when the wood is burned is the response. The polar relation of projective geometry is here realized by Nature on a gigantic scale, but the one element of the dual process can only be grasped with scientific thinking when the idea of ethereal or negative space has been developed.

To understand what is here intended, it is essential to unburden the mind of atomistic preconceptions, not with a view to their elimination, but in order that the visible and tangible phenomena may tell what they have to tell from another aspect. By far the greater part of our atomistic world-picture is not phenomenon at all and cannot be; it is idea. This makes it no less real if the idea is true, but it should not become a "fixed idea", barring the way to others which the intuitive mind may also read from the phenomena directly, and which may prove no less essential. The planar realm when found will complement the pointwise; the one idea will help interpret what is enigmatic in the other.

It is in physical space and above all in inorganic matter – or matter which, like burning wood or coal, has fallen out of the living process – that those expansive, dissipating tendencies are manifest which are at last summed up in the Second Law of Thermodynamics. A classical example – the irreversible diffusion of gases, each of them tending to fill the available space till its own partial pressure is balanced – shows how the Second Law is connected with the expansive, space-filling tendencies which are evoked by heat, and therefore by radiant energy as such, in inorganic matter. But in the realm of life, if we are unbiassed – if observation and thinking are truly poised between the two "dual" possibilities which we now know to be inherent in the very nature of space – we shall associate radiant energy with a centripetal and not *only* with a centrifugal tendency. Seeing how life renews the differentiated forms and substances which death disperses, we shall not unduly universalize the Second Law.

The concept of ethereal space makes it possible to conceive in a simple way both the inward and the outward tendencies of radiation. (See again the top right-hand picture in Plate X, supposing the apparent "source" to be in the region of the point-at-infinity within.) The outward tendency is like a primary phenomenon of levitation, evoking once again the kinship of "Leichte" – "light" and "lightness" – suggested by the genius of language. The plane of levity is the plane-at-infinity of physical space. The inward on the other hand, of which we have been speaking, is a space-filling tendency in negative space, working from the periphery towards a boundless realm within, which in comparison to matter-filled space is like a space that is "more than empty". The outward and inward tendencies balance one-another; the space is filled with "ethereal light", which may well be of many kinds; the dynamic balance here suggested may be established in many different ways.

A "space of light", thus conceived, is indeed like the "negative" of air – of the Earth's atmosphere which finds its balance between the centrifugal, space-filling tendency of gases, and the centripetal of its own weight. The interplay is manifest in atmospheric pressure; the balance is established in a gradient of pressure. So in the corresponding realm of ethereal space and substance, the interplay will manifest in an ethereal or planar field of force for which the proper word is "suction". Unlike most "suctions" that occur in physical mechanics, this is, however, a true, qualitative force of suction, not the mere outcome of a difference of pressures. Be it observed that in this polar comparison of "light" and "air", the centripetal tendency of the one corresponds to the centrifugal of the other and vice versa.

All these things indicate that in radiation there is a *primary* activity peripheral in nature, tending inward and not only outward. The physical centrifugal effects are in the nature of a polar-opposite response. Once again the relation of centre and periphery is mutual.

The implications for physics and chemistry are far-reaching. The numerical rhythms and harmonies shown in the spectroscopic phenomena, deeply related as they are to the laws of chemical affinity and structure, will have their origin not only in a sub-microscopic world of atoms, but in this interplay of centric and peripheral activities. As in projective geometry the pure relations of number working in the ideal interplay of point, line and plane are found to govern the enhancement of ever more complex forms, so will the primary rhythms, manifested in the differentiations of matter and radiant energy, be working in the cosmic realm of Light and Dark – using the Goethean terms in their wider meaning. If then the spectral lines are an indication of these cosmic rhythms, antecedent to matter though embodied in it, the discovery of the same lines in the light of Sun and stars does not oblige us to infer that matter, ready-made and precipitated into Euclidean space, is present yonder as it is on Earth. What it makes certain is that the formative forces to which the earthly matter is due, are also there in the great universe.

Thinking of Sun and Earth, and of radiation in its relation to matter, we have approached the Goethean polarity of *Light and Dark*. Goethe meant not merely visible light; the purest and most incandescent white was for him already an outcome of the interplay of light and darkness.[47] Moreover the dark was for him associated not only with unsubstantial shadows, but with the very essence of the matter which throws the shadows. In "Light and Darkness", he divined an ideal polarity, fundamental to the structure of the world; the visible phenomena were an outcome of the interplay of both components. This does not mean ultimate dualism, any more than does the geometrical Principle of Duality. The more one enters into Goethe's concept of polarity, the more deeply does one see it to be related to the projective geometry which in his lifetime was only just beginning. Today, when his scientific work – including the *Theory of Colour* – is being taken more seriously than in recent generations, one cannot but confess that his idea of the realm of light, near as it is to our immediate experience of Nature, is foreign to the existing theories, whether in their corpuscular or in their undulatory aspect. The discrepancy may be resolved when it is seen that these theories refer to the *physically spatial* effects of light – effects which are only brought about when light is already caught and held, so to speak, in darkness.[48]

Using the Goethean ideal concepts, physical or pointwise space as such belongs, in effect, to the "dark" pole, and negative or planar space to the "light". Yet the polarities of the world are so interwoven that the former is sustained by an unique plane – an unique entity of light – and the latter by an unique point or entity of darkness. Formative, crystal-begetting light from the infinite plane "informs" the Euclidean spatial world, the space of darkness. And in the dark fertile earth every living seed or germinating centre is the receptive focus for a vast space of light – the planar space which bears the archetype of the organic form that is here seeking realization.

We have thus approached the phenomenon of radiation from two polar aspects. Let us think again of the typical and familiar example, – a candle flame! We contemplate it and think as follows: It radiates light, heat and chemical effects. In this process, not *only* the radiation from within outward is at work; for the candle *can* only radiate outward, because in the ether-space, of which the all-relating point is *in* the candle flame, the ethereal forces work inward from the world-periphery. The flame is a *receptive* centre: only thus can it radiate, bestowing light and warmth. Centre and periphery work together in active and mutual interplay.

The radiation in this case is connected with a chemical process. If our hypothesis holds true, we can gain through it a new insight into this process. The chemical interchange consists not just in the moving around of pointwise (atomistic) particles in physical space, but in a process of becoming and passing away again (*Werden und Entwerden*). The substance which is burning – now in space – originally came into being from the periphery. It has come to rest in a quite definite, rhythmically determined, qualitative-quantitative equilibrium between centre and periphery. Now, through the chemical process engendered in the flame, an inroad has been made into this restful state. The substance is being "eaten up", and a new equilibrium is coming about in the new substance which is being engendered.

Heat in its very nature is always delicately in the balance, in the interplay of ethereal and physical spaces; it is at the threshold between the ponderable and the imponderable or "antiponderable". The view held by the nineteenth-century scientists, as expressed in the Theory of Gases, came nicely to expression in the famous, classical textbook of the English physicist Tyndall: *Heat as a Mode of Motion*. Rudolf Steiner answered this in the following words: "Yes, it *is* movement, but not outward, spatial extensive movement – not just a hopping about of the molecules, and so on – but *intensive movement*,[49] whereby is meant an inpouring of the peripheral into the centric, and a working outward of the centric into the periphery; an interplay of physical and ethereal spaces."

The idea that in all processes of radiation peripheral space plays a part is to be grasped in this sense. Underlying the active process, which appears to come from a radiating *centre*, is the equally important part played by the universal forces, which work in from the periphery. Yes, the original source or origin is to be sought in the periphery rather than in the centre.

Chapter V

ETHEREAL SPACE OF THE PLANT SHOOT

38   Convex and Concave Growth

In physical construction, dominated as it is by point-centred concepts and measurement from material centres outward, the relationship between *convex* and *concave* is a purely external, spatial one, recognized from the physical point of view. Here the most fundamental experience is architectural. In earthly masonry, we must build brick on brick, add stone to stone, each of them impenetrable, external to the other.

The convexity of a building, and the empty, concave aspect of its interior, reflect purely and simply the nature of physical spaces and forms; there is an inside and an outside. The larger the building, the greater its potential content.

Architecture takes on many forms, and where the attempt is made to create surfaces or plastic shapes by means of planes, rather than as the locus of a moving point, the impression may be evoked of an enveloping and enfolding quality of form, which immediately begins to lift the architecture out of a mere relationship with centres of gravity; something else begins to speak through the forms. It then becomes unsatisfactory simply to set buildings side by side. The question of a qualitative, mutual relationship *between* the buildings, and between their outer and inner aspects begins to require quite other answers. Herein, indeed, lies the future for the art of architecture.[50]

In contemplating living organisms – in particular, for example, the forms of the higher plants – the gesture is revealed of *planar* forms, primary and immediate. It is shown often by a single leaf, often by several or even many hundreds of leaves together, as in the beech-woods in early spring. Here, Nature, we suggest, is placing before our eyes the kind of space in which the plane is primary and not the point. And such a space will be endowed with a more definite orientation, form and measure, if there is somewhere an innermost point acting as the "infinitude within" of a living, *concave* space; just as the outermost plane or infinite periphery gives form and measure to the space of Euclid, and to the forms characteristic of Earth-space. *For each growing shoot of a plant, there will be such a point at the focus of the typical hollow space above the apex of the stem, and to its form-giving presence the enfolding leaves bear witness.* We shall refer to this as the Sun or Star-centre.

Once more, we will formulate our fundamental thought:

The form-giving life of Nature is determined not only in the Euclidean universal space in which matter as such, qua earthly matter, is at home, but also in the polar-opposite type of space-formation. It is a *type* of space-formation, not a single universal space given once for all. Spaces of this type come into being and pass away again with the life-cycles of living creatures and of their several organs. Wherever in effect there is a living seed, a germinating point, a special focus of life or growth – whether within the watery substance of a living body or hovering outside it as the growing-point of the higher plant – there we shall look for the "infinitude within" of such a space-formation. And as this space-formation is of the *planar* type, we shall find evidence thereof in the plastic formative activity around the point in question, or in the gesture of the leaf-like organs that envelop it and thence unfold.

Applied to the understanding of Nature, the concept of ethereal spaces involves a two-fold approach. First, the *morphological gesture* of the living entity must speak, so that it shows how the ethereal spaces are formed, where the ethereal centres or points-at-infinity are located within the living, developing form, and how their fields of action are interwoven both in time and space.

The relative median planes and planes-of-levity will need to be recognized, as well as the more obvious geometrical features, such for example as concentric spheres or again spiral or polyhedral forms, many of which will appear rather different in the ethereal than in the physically spatial aspect, so that some regularities of form and pattern will only be recognizable to an awakened eye. (Consider, for example, how the conception of the "ethereal concentric spheres" was reached, as pictured in Plate IX.) This aspect of the problem will be mainly a question of morphological insight and geometrical imagination acquired through practice.

The second range of problems is more difficult. It is to understand just how the *physical-material substance* of the living body is received into the field of action of the ethereal formative forces. How, in the processes of life and growth, do the physical and ethereal interlock and interweave? To what extent will it still be true to say that in their purely material aspect the processes in the living body are held in balance by physico-chemical forces as hitherto understood? Or will the insight we are now deriving from the living form ultimately change our outlook on these forces too, so that in time to come we shall learn to put the question in quite a different way?

These are the two directions of our problem. Taking them both to some extent together, our line of thought is as follows.

The simplest manifestation of life will be a growing watery or semi-fluid sphere as in a single living cell, a unicellular organism, or the morula and blastula of animal embryology. Life here reveals itself in convex, outward growth, nor should we expect this to be otherwise. For the living body is material, and a material body is *ipso facto* more or less filled with matter from within outward, presenting, on the whole, a convex surface to the outer world.

The living sphere at this stage differs from the dead not so much in its form as in its growth and other living functions. We see it now as the manifestation of an ethereal space, with the point-at-infinity within, in the watery-living substance. This will be the *ethereal focus* of the space, or of the form in question.

The primitive phenomenon of life and growth will always be of this kind, as in the simplest instance when a spherical or spheroidal organism, with the point-at-infinity at its centre, lives and grows at the expense of its surroundings. Growth then appears as a primary phenomenon of suctional, levitational expansion. In such a case we shall speak of *primitive* or *convex growth*.

In life and growth, the physical material, otherwise only subject to gravitational and other physical forces, is in some way linked on to and received into the ethereal, formative activity. In formative tendency and function it becomes more surface-like; greater surface-areas are formed, where without life it would remain predominantly point-like, inert and centred within its dead, material substance, subject to disintegration and decay – "dust unto dust".

It is essential to our hypothesis to conceive that it is the watery matter of the Earth, which, once received into a living organism, is capable of this change – this surrender to the ethereal, peripheral qualities of space-formation.

Here, indeed, we touch on the whole question of the properties of all fluids, and of water itself in Nature's great laboratory. We refer the reader here in particular to the writings and researches of Theodor Schwenk.[51] The phenomena of *sensitive surfaces* in flowing water (in air and warmth also), and of *surface tension*, demand renewed insight in the light of the concept of planes of levity or Leichte. So also new ideas must be brought to bear on the phenomenon of osmosis, which in the non-living world follows a fairly simple course; in the living organism it will take on a new light, once the concept of Leichte has been assimilated into the science of biology.

There is already, in modern biochemistry, a tendency to overcome the old additive habit of thought which would regard the organism as a mere summation of the molecular events going on in each single cell or volume-element within a cell. Biologists are amazed by the subtle and pervasive quality of the chemical influences – hormones and auxins for example – with which the living organism regulates its life and growth; also by its selective power with regard to osmotic and ionic processes.

From the whole concept here developed – the interplay of centric and peripheral formations and forces – it may well be expected that living forms, seen as a whole and penetrated with this kind of understanding, will also provide essential clues to the nature of the substances which they produce and to the chemical formative processes which are today interpreted as an outcome of the minute molecular constitution.

Looking once more at the primitive living cell or sphere and thinking of it as an ethereal or planar space-formation with the ethereal focus somewhere within it, the orientation of the ideal tangent planes will in this simplest instance be in relatively concentric spheres about this focus (Plate X top right). Predominant, in the phase of growth, will be the ethereal force of levity. The living sphere expands, taking in watery substance from its surroundings. This is the primary phenomenon of vegetative growth. It is important to recognize that it is primary – as Goethe would have said, an *Ur*-phenomenon. We have no elaborate knowledge of its technique; we recognize it as an elemental fact.

For an analogy, polar opposite in kind, imagine the physical planet Earth – the forces of gravity there prevailing. All over the surface of the Earth – in mountain lands where avalanches fall and rocks are loosened; in hail and rainfall; in ocean breakers the falling of the spray; in myriad variety of accident and incident – material bodies are ever falling, down towards the centre of the Earth. This too is the "Urphenomenon," the geometrical lines of force and dynamic character of which we can clearly recognize, without indulging in any theories of hidden cause. In like manner we can recognize the Urphenomenon of living growth as a polar opposite phenomenon; it is one of planar, plastic, peripheral and suctional expansion. The planar forces *function peripherally*, just as in the inorganic realm the centres of gravity function centrically. If the primitive organism dies, its substance is indeed disintegrated and completely abandoned to the inorganic realm of natural substances; while it lives, it is held in balance between the cosmic and the earthly realm.

Applying this concept of ethereal space-formation to living organisms, and guided in so doing by the phenomena, we recognize at a very early stage an essential differentiation. Besides the type of organism, where the point-at-infinity of the ethereal space is immersed in the watery, living substance, there is another basic type, where the ethereal focus may be outside the confines of the organism, poised in the open air or in the surrounding medium.

It is evident from all but the simplest of morphologies that the living form is determined not only by convex outward growth but by another kind in addition – by an ethereal space which, though its focus, to begin with at any rate, is outside the living body, yet belongs organically to it, determining the way it grows. The growing body produces organs or tissues tending to enfold and envelop the ethereal focus in question, thus making evident the presence of the latter. We then have the phenomenon of *concave growth*. It will be evident enough that primitive or convex growth must be there as a foundation before concave growth can ensue. The interplay of convex growth and concave leads in the simplest instance to a bipolar living organism. Axial development is therefore often to be interpreted by the presence, at either end, of an ethereal focus of the one and of the other kind. Precisely this is what we recognize in the higher plant, root and shoot (Plate III, Fig. 14).

Of the two types of growth the convex is the more earthly; here the ethereal is more immersed in and surrendered to the material and physical, and the resulting organic forms, though they are living, will often adapt themselves to the more centric and radial structure of the latter realm. The concave on the other hand is related to the Sun and to the realm of light, where the ethereal is manifest more fully in its own domain, unalloyed by matter. For this reason we described the focus hovering above the growing-point of the shoot as a sun- or star-centre, and the space which it engenders in the formative organism of the plant as sunlike. The word "star" is used inasmuch as the Sun is a star among stars. As there are countless individual stars, so are there many species of flowering plant. The "star" above the growing point determines a sunlike space in every instance, but this may be of diverse qualities, just as a physical space, one in type, may be diversely formed and filled with substance.

Thus the polarity of Sun and Earth occurs again, microcosmically, in the living world. Over against the sun- or star-centre, the focus of convex growth with its more earthly character may also be described as the relatively "dark" pole. It is once more the Goethean polarity of light and darkness, but with the light working within the darkness at the convex pole, while at the concave the darkness reaches out to embrace the light.

39   *Animal and Plant – Invagination, Evagination*

In the phenomenon of concave growth a further great differentiation is recognizable. The living organism, growing initially as from its convex centre, can respond in either of two ways to the additional presence of an ethereal centre which, to begin with, is outside and yet related to it. It will produce, as we have said, organs that envelop and enfold this second centre, expressing the peripheral, tangential nature of the ethereal space which it determines. But this enfolding can thenceforth proceed in either of two directions. In the one case the living organism draws the more sun-like ethereal space ever more deeply into itself. Beginning with a very slight concavity on the surface of the formerly convex living sphere, the enfolding forms tend to embrace the new ethereal centre ever more closely; they grow up and coalesce around it, until at last a hollow sphere is produced, more or less wholly immersed in the body. This is what happens in animal development; it is the typical phenomenon of gastrulation – invagination. In the play of dimensions the ethereal infinitudes may also take on a linewise form; such a *focal line* of concave growth underlies for instance the development of the neural groove, leading eventually to the spinal cord. The further study of animal development from this aspect is beyond the scope of the present work.

The other alternative predominates in the plant kingdom. The organs, namely, of the living body which grow outward to enfold the "star-centre", instead of tending to embrace it ever more closely and thus draw it inward, thenceforward open and expand, in spherical or conical tangential forms which in their further growth reveal the typical geometry of spheres or cones in negative or planar space. In fact the typical unfolding gesture of the shoot, opening the hollow space which it originally formed at the focus of concave growth, might rather be described as a perpetual *evagination*, contrasting with the repeated invaginations by which the finished animal body is produced[52] (see illustrations in Chapter I). In the growing shoot, the star-centre is not claimed nor drawn down into the dark watery body; it is left free. Ever new leaf-like organs are produced, closely enveloping it to begin with and then opening away. It thus becomes an inexhaustible source of fresh life and growth. This is the typical phenomenon at the growing-point, described in Chapter I.

40   Matter as a Receptive Matrix – Chaos and Cosmos

The differentiation of convex growth and concave, important as it is, is not as absolute as might at first appear. For an ethereal space – compare the pictures of Plate X or the linewise curves in Figs. 31 and 33 – is by its very nature *concave*, tending towards a potential emptiness at the infinitude within. If an ethereal field is to work at all, it must be working in this way even when the focus is immersed in matter. It can only "lighten" and enliven matter by overcoming the inertia whereby each heavy particle maintains its own centric position – stays where it is until pushed or pulled by others. We do not pretend to have solved the important question, how the ethereal field is "hitched on" to the material. But as a general principle it is clear that in the region of an ethereal infinitude the matter must in some degree surrender its once-established molecular or pointwise structure and become "chaotic", like an unwritten slate – a receptive matrix, a *materia* in the original sense of the word. As regards form; a kind of void must be there at the ethereal infinitude, to enable the ethereal formative forces to engender form anew. Though the suggestion that this will happen even at the molecular or atomic level may appear revolutionary, something of this kind is not unknown to biologists; namely where fresh developmental processes or far-reaching metamorphoses are to take place, tissues and cells already formed often return to a more embryonic, meristematic, relatively formless condition. An outstanding example of this "chaoticizing" process is the butterfly's pupation.[53]

Referring to the realm of the ethereal focus or "infinitude within", we are here using the word "chaos" in contrast to "cosmos" – manifest and ordered form – rather in the ancient sense. "Chaos" or "matrix" indicates a realm where some specific form is either not yet in being or where it loses hold and passes over into a relatively formless state. Matter that holds fast to its given structure – molecular, cyto- or histological or on whatever level – will be to that extent unable to surrender itself to the formative forces of living growth; it maintains its form, remains, in other words, a "cosmos".

It is of the essence of a germinating cell, that within it, in the region of the ethereal focus, there will be something of this quality of *chaos*. The "infinitude within" bears the *archetype* of the future form. It is the form-giving *concept* which is in some sense focused in the "infinitude within"; it cannot in the nature of the case be already manifest therein. It *becomes* manifest towards the surrounding surface or surfaces. Hence it is symptomatic that primitive organisms are in so many instances most clearly and elaborately formed at the periphery, even with a siliceous or calcareous exoskeleton, while the comparatively unformed portion is within. The section through the growing-point of a plant also gives an unmistakable example of this peripheral mode of formation.

Once more, the polar opposite analogy may help us. We have already mentioned (paragraph 26) how in the light of projective geometry the crystal lattice – characteristic form in the mineral kingdom – is seen to derive from an archetype in the infinitely distant plane or ultimate periphery of space. In the nature of the case, the form cannot be manifest in yonder plane. The archetype is unmanifest in the periphery; the crystal form becomes manifest about a centre, somewhere in earthly space.

To dissolve a crystal of some salt so that every trace is lost beyond recovery, we shall dissolve it in the greatest available volume of water. The salt diffuses out towards the infinite sphere, the very realm from whence ideally the form derives. This is indeed the way we treat – or the way Nature treats – every no longer wanted form: it is scattered far and wide, into the void, and the great void receives it in beneficent oblivion. And yet the uttermost extremes of yonder void bear the ideal archetypes of form – with the imaginary circle the very archetype of space itself, also the individual formative types of all the crystals.

This therefore is the essence of physical matter, in so far as matter belongs to the inorganic world, filling the space of Euclid: chaos and void in the periphery, manifest form within, in and about each given centre. The opposite will be true of life, inasmuch as the phenomena of life belong to a negative-Euclidean type of space-formation: there will be relative chaos within, at the very centre from which materially the life appears to proceed; manifest form as we go out from thence.

The "going outward" involves the time-process of growth. Primitive microscopic forms as we have said, like some of Haeckel's protozoa with their reticulate and polyhedral exoskeletons, place the idea before us in a symptomatic, purely spatial picture. For every higher and more complicated organism, growth begins from a minute germ-cell. What afterwards appears, vastly increased in size, in powerful and manifold formation, began in a minute, point-like region. Here, we suggest, the opposite of the customary physical way of thinking is applicable. We have not to look only for the molecular pre-formation of a complicated organism inside this tiny space. We must also apply the opposite idea. Here is the "infinitude within" (corresponding to what in physical space would be the infinite surrounding void, the empty nothing). The formative space, to which the life and growth are due, *weaves around it*.

In this direction we should seek the explanation both of specific living form – which as a rule begins with a spherical and plastic outline, and in which surfaces and skin-formations are all-important – and of the driving force of life and growth, which appears "suctional" in character. This way of thinking, we believe, will contain the key to many typical phenomena, well-known to biologists: the character of embryonic and meristematic tissues, also what happens in the living organism when at a later stage of the creature's life new vegetative growth is needed or else a radical metamorphosis is about to set in, as in the transition from the caterpillar via the chrysalis to the imago, or in the leap from leaf to floral organ.

Thus the phenomena of Nature wherein the forming of ether-spaces plays a predominant part differ from the physical and inorganic, the rationale of which is contained in the geometry and mechanics of the space of Euclid. The difference is so far-reaching as to suggest a certain inversion not only of space but of time itself – an "experiment with time", to borrow the late Mr. J. W. Dunne's well-known title.[54] It is at least a valuable exercise in thought, once we have grasped the fundamental notion of negative or ethereal space-formation, to follow the growth and development of a living creature backward in time from the mature organism to the germ-cell. In this return from highly articulate form to relative formlessness and chaos – remembering that for the negative space the approach to the ultimate point or infinitude within represents "*inward increase*", we are reminded of what is constantly happening in the inorganic world, but now in positive space and forward-going time. Well-ordered forms that were occupying comparatively small spaces tend to disperse and to chaoticize. Enlargement goes hand in hand with loss of form and ordered pattern.

This again sheds a new light upon a well-known chapter of modern science, of which the philosophical and cosmological implications are far-reaching and very difficult, namely the famous second law of thermodynamics. Since Maxwell and Boltzmann this is understood as the expression of a chaoticizing process, a progressive loss of differentiated form and on the whole a tendency of dispersal as when different gases for example tend to mix, each of them wanting to occupy the whole of the available space once it is free to do so. It is an "irreversible" process.

All the conceptions of probability which modern physics attaches to these phenomena begin by taking for granted a Euclidean type of space, inhabited by molecules or atoms of a quasi-physical kind. Hence the inevitability of which Sir Arthur Eddington so eloquently wrote with regard to this "second law". Yet to return to the phenomena themselves: if we have watched the potent growth of a living organism, making its form, so to speak, out of formless water, do we not also gain the impression of a certain inevitability in this silent power? Reverse the flow of time and simultaneously invert the space so that the ideas of large and small are interchanged in the precise geometrical sense we have explained: the one becomes reminiscent of the other. If time went backward, living organisms would as inevitably appear to lose their form towards the germinating point from which they spring, as do the distributions of inorganic matter scattering to formless mixture. But as things are, in the forward flow of time the two tendencies are mingled, and life renews the forms which death disperses.

Thus if the deep polarity of ideal space proves applicable to the phenomena of Nature as we are here contending, a new page will be opened in the history of science, also in this respect. But the further influence of these ideas in physical science generally is beyond our present scope.[55] It is the living world which is opening our eyes in the new direction and we must tarry longer in this school, learning to walk before we run.

We have described the fundamental notion, reading the primary phenomenon of living growth in terms of ethereal rather than physical space. But the material which is thus formed is physical to begin with; moreover the outcome of the living process is again and again to secrete and deposit more or less hardening materials in shapes that serve the living form and bear its imprint. To the extent that such material, even if still within the living body, becomes a finished form and is no longer imbued with the full force of vegetative life, it is again more or less paramountly subject to the physically spatial world.

Thus the two forms of space and force are interwoven; probably neither is without the other anywhere, but we can recognize where the one or the other is predominant – the one in germinating regions of fresh life, the other where the living substance becomes hardened to a mere supporting organ, and even mineralized, thrown out of the living process. Whether a modicum of ethereal activity permeates even the so-called "inorganic" world is a question we may here leave open. To understand the living world we must first apprehend, clearly and radically, the notion of ethereal space and of the forces and formative tendencies which this involves, and we must then apply this notion not to the exclusion of the physical but in conjunction with it. To find the interplay and balance of the two will be the test of insight.

41   Sun-space at the Growing-point

Returning now to the morphology of the higher plant, we come back to the phenomena described in Chapter I. There at the tip of the growing shoot is a hollow space, often deeply hidden and protected amid the young enfolding leaves. It is in the nature of the case, as we have said, that there must first be convex growth to a certain extent, before the typical concave hollow gesture can appear. The archetypal phenomenon of concave growth at the growing-point is preceded at the microscopic level by convex growth; the budding leaf-primordia, meristematic growth-forms at the growth-cone of the shoot, as seen under the microscope, are recognizably convex; but as the organs develop and become visible to the naked eye, the forms reveal more and more a concave gesture (Fig. 59). Within the hollow space enveloped by the young leaves, there is the all-relating point, the inner infinitude of a sun-space. The plant develops on a small scale its own inner sun-space, or spaces, which are related to the macrocosmic, heavenly sun-space.

The focus of this ethereal space is within the hollow, which in the higher plant is always there, *above* the material growing-point of the stem, its presence indicated by the enveloping gesture of the young leaves. We see a spherical, spheroidal or conical form of growth – a realm *not filled with substance*. The ethereal focus hovers above the actual substance of the growing stem, and the sphere, which the young leaves envelop, as they expand and come away, is like the hollow of a chalice, open to the light-filled air. We see it, for example in the Dandelion in Plate XII.

In cultivated plants like the Lettuce or the Cabbage, the phenomenon of the hollow space above the tip of the stem is often very evident, especially if the plant

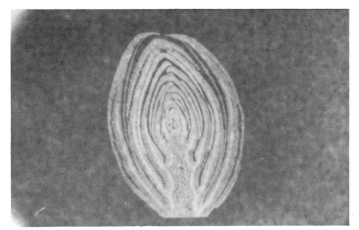

59a   *Woody Bud*

Space of the Growing Point

59b   *Coleus*

59c   *Coleus*

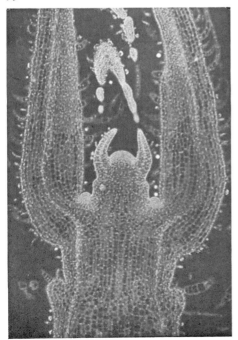

59d   *Elodea*

begins to "bolt" (Fig. 60 and the Brussels Sprout in Plate III). Cultivated as these plants are, in order to develop an accumulation of nutritive leaf-substance around the vital growing-point, the hollow space has become densely crowded with young leaves. A well-grown lettuce or cabbage, with a good "heart", presents a round and firmly filled convex ball. As soon, however, as the heart begins to move and open up in further growth, a vertical cut will reveal clearly the hollow space within, created by the very young leaves. In this space is the sun- or star-centre – the heart of the growing plant or of the shoot in question. It is a phenomenon so entirely opposite to what can be seen at the root of the plant, which grows downward into the soil.

At the tip of the upward-growing shoot is a realm which is moulded and plasticized by the planar surfaces of the leaves. It is the very gesture of these planar surfaces which announces clearly where the ethereal focus – the sun-centre – is poised. To the awakened observer, the exact situation of the ethereal focus is clearly to be seen, in the spatial gesture of this hollow realm. Sometimes, especially in the early stages, the hollow space is very *spherical*; later on, and often from the outset, it is a *conical* hollow which is typical. The *type* of gesture is shown in the top right-hand picture of Plate X, in Plate IX and in the "cone-space" of Plate IV. A horizontal section through a growing-point will reveal the beautiful peripheral arrangement, often in spiral form, of the leaf-primordia or of the young leaves, which together create the sun-space (Fig. 52).

In the more conical gesture, the hollow cone-space comes to expression in the greater Plantain (Plate V). Here, the point-at-infinity of the space will be somewhere on the vertical axis, for the axis is the *line-at-infinity within* of an intensive, two-dimensional space. (A line-at-infinity of an ethereal space will be *in* the "all-relating point" of the space, just as in the physical space a line-at-infinity will be in the "all-embracing plane".)

The conical gesture is in fact the most general form in which the plants reveal the counter-spatial nature of the realm from which they grow. As we have learnt in earlier chapters, we must allow the *planar* aspect of formative processes to speak through the phenomena, as well as the point-wise. What speaks here so eloquently is that the leaves – *planar organs held in a point, the node* – create the space within which the younger organs will be formed. Looked at from the aspect of substance, the process is one of gradual densification and hardening. A woody node and tough, fully-grown leaves are secondary, not primary. Primary are the young, delicate growths, which, reaching upward with a more vertical gesture to envelop the sun-centre above the growing-point, thus forming the ethereal cone, tend out and away from this inner region towards the horizontal. It is an archetypal phenomenon revealed in multitudinous and manifold ways in the plant world. Even where at first sight this gesture would appear to be absent, a closer look will often identify it. To take two common examples: in young Rose shoots, the leaves are folded flat, and the whole process takes place at first in a plane, which later opens up like the leaves of a book to reveal the hollow. In the Sycamore or the Begonia, only the first tiny leaves are vertical and there is a very speedy, though eloquent, opening out into a most beautiful horizontal plane,

61b   Begonia Shoot

60   Section through a Cabbage

61a   Sycamore Shoot

62   Grasses

after which the leaves, as they grow older, quite often become strongly convex over the whole or most of what has now become their upper surface (Fig. 61). The transition from concave to convex in the development of leaves is of primal significance in the language of plant morphology; it is an archetypal phenomenon revealed by the higher plants. In so many different ways in the different plants, the leaf, at first a concave surface, hollowed inward towards the point or line-at-infinity of an ethereal space, makes the transition through a more or less horizontal plane to a more or less convex form, which then tends to hang down below the horizontal towards the Earth. See for example the Rhododendron in Plates II and XX, or the Dock in Plate VIII.

It is in the delicate and gentle interplay between concave and convex surfaces that the plants reveal, through their green leaves, their essentially ethereal nature. The green leaf in full balance between the extremes of unfoldment often reveals in the most breath-taking way in spring and early summer, the buoyant levity or Leichte of the ethereal planes. The Sycamore and the Beech, for example, each do this to perfection, but in quite different ways. The Hazel twig in Plate I shows the typical transition from concave to planar surface. In the Rhododendron in Plate II the transition from concave to convex is revealed by the leaves of the nodes of successive years; we see a record of the development of the branch through three years, and within the delicate, yellow-green cone of young leaves is to be found the flower-bud of the coming year. The leaves of the two lower nodes will have formed just such a cone in previous years, and in their turn will have opened out through the horizontal, becoming convex, much tougher and of a dark green colour, as they turn their upper surfaces with a strong convex gesture towards the rays of the outer sun. Below, in the dark recesses of the woody parts of the bush, it will be possible to find the old brown leaves of still earlier years, hanging on, before loosening their hold and falling to the ground.

The characteristic gesture is to be seen in manifold variety in the growing shoots of the plants. The growing tip is always hidden amid the tiny budding leaves, which as they grow reach up above it as if with protecting hands to guard a hidden treasure. Yet it is an empty space they hold between them. Perhaps, as in *Silphium* Fig. 11, they are ranged around it in pairs – see also *Hydrangea* and Rowan (*Sorbus aucuparia*) in Figs. 1 and 2 – perhaps in whorls as in the Woodruff (*Asperula odorata*) in Fig. 7, perhaps a single leaf with its concave inner surface tends this hollow space as in Blackberry (*Rubus fruticosus*) in Fig. 6. However it may be, the concave hollowing of the inner surface of the younger leaves around the space above the growing-point is characteristic. Each single shoot holds out its planar surfaces as though to receive its portion of the light.

With further growth the leaves expand; between them other, younger buds grow to take their place, while they with increasing maturity open out more and more towards the horizontal. There, plane upon plane, they are outspread high above the surface of the earth, organs to receive the light and air, abounding with the buoyant life of springtime.

V

Even the slender grasses envelop a space in the centre of their growth around which their leaves are at first wrapped and folded and then open out. From it, later on, the flower and the seed will spring; the golden ear of corn has such a birthplace. The lower part of each leaf (the petiole) forms a sheath close around the stem, and around the petioles of other leaves, each creating the sheath, within which the next one arises. It is from this "infinitude within" – the heart of the grass-halm – that the ear of corn springs, which can be found in rudimentary form, by peeling off the outer layers at a very early stage of the plant's growth (Fig. 62).

Even an individual leaf, once it has opened out and expanded into a visible form in space, usually retains for the whole of its life some aspect of the hollow ethereal space out of which it was born. There is usually a slight hollow at least along the mid-rib. Many typical, fully developed leaves reveal the interplay of concave and convex in the saddle-form; the surface, when followed from one leaf-edge to the other, remains concave, while the mid-rib may curve backward from petiole to leaf-tip in a strongly convex gesture. This saddle- or paraboloidal-form is typical of many leaves as they grow older; prime examples are the leaves of apples and pears.

The very surface of the leaf itself, in its delicate and varied balance between concave and convex is thus a clear indication of its organic role in the rhythmic system of the threefold plant, poised between the extreme forms of root and flower.

Let us look again at the Greater Plantain (*Plantago major*) in Plate V. Its young leaves enclose a conical hollow. Sheltered in the space in the centre of the plant, each leaf reveals this enveloping gesture. The conical spaces which are enveloped by the very young leaves are deep and vertical and tend toward the central axis of the plant. Low down at the base of each leaf-stem or petiole is the bud from which the flower-bearing stem springs, growing vertically upward towards the light. Usually the young flowering stem begins by nestling in the hollow groove on the inner side of the petiole. It is as though the whole leaf were formed so as to shelter a flower-bearing stem and give it a space in which to grow. With each successive leaf, springing from within the heart of the plant, this enveloping gesture is repeated, but as the leaves grow older and larger, they gradually lose their hollow form and the space disappears. The leaves open outward and become flat or convex until at last they lie on the ground and die.

The young Beetroot plant (*Beta maritima*) shows this gesture beautifully (Fig. 14). Here again the form is of a cone opening upward; the leaves surround the as yet invisible central axis of the plant. The younger the leaves, the more vertical is their posture; but they will open and spread out until at last they die upon the surface of the soil and give their substance to the Earth. Below is a form of quite another character, the rounded, swollen beet, with its tapering root. This part, about which we shall have more to say later on, is of an earthy nature, packed with material substance.

It is essential to be clear about the true nature of the "cone-space" into which the Greater Plantain has been drawn. We are taking it as an ideal thought-form to describe the *negative space according to the laws of which the plant shoot appears in positive space*. Let us describe once more this family of cones in a point (Plate IV).

The geometrical picture of a family of circular cones in a point must be seen as a negative space-form, and not as though it is three-dimensional in positive space. The cones are *ideal* forms. The accurate geometrical description of them, we have said, is that they are *intensive* one-dimensional forms within the *intensive* two-dimensional space of the point which bears them.[20] Although these cones are not to be thought of mathematically as three-dimensional forms, yet the interesting fact is that *their gesture of form* may be recognized throughout the plant kingdom in its developing vegetative organs. The cones portray a process taking place in ethereal space; they represent an ethereal space-formation. These cones must be thought of as being formed *planewise*; they are enveloped plastically by all their tangent planes, which are held in the *one point* and stretch away to infinity on all sides. The cones open upward and downward to infinity. Held in one point, they are "concentric" between the vertical line (pink) and the horizontal plane (green), which must also be thought of as extending on all sides to the infinitely distant horizon. As the cones open out, one after the other, they spring never endingly from the inner axis; they widen, becoming ever flatter and at last disappear into the horizontal middle plane.

This family of "concentric cones" (or better said, "co-planar cones") is in the exact geometrical sense polar to the family of pointwise circles in Plate X (red, top left). Just as the circles move inward toward their central point and outward towards the infinite circle, which is the infinitely distant line of their plane, so the cones open out towards the horizontal plane and close in towards their vertical axis, which *in ethereal space* plays the part of an "infinitude within". The concentration of the red circles in Plate X towards their centre, which we think of as a contraction, corresponds to the opening out of the hollow cone-forms. If we think of the contraction of the circles towards their central point as picturing a process in physical space, then we must think of the opening of the planewise cones towards their middle plane as picturing *a process of contraction taking place in counterspace* and vice versa. In Plate IV, the positive-space circles have been drawn in green on the horizontal plane of the cone-space. If the circles are radial, filled pointwise from their centre outward, the cones are peripheral, empty and hollow inward towards their innermost axis.

A most beautiful plant to study, when we try to familiarize ourselves with this method of observation, which opens up a whole new approach to plant morphology, is the Dandelion (*Taraxacum officinale*) as depicted in Plate XII, and in all its forms. A dandelion seedling, which has germinated with freedom of space, perhaps in cultivated soil, often shows with remarkable clarity a conical or a bowl-shaped form. Every leaf contributes to the moulding of a perfect bowl or hollow cone, which simply opens out as the plant develops, until one day a green flower-bud appears, nestling at the bottom close to and maybe even below the surface of the ground, so deeply hollow is the green bowl in which it appears. In the ensuing development, the bud is raised on its lengthening stalk and presently the golden yellow flower opens like a sun poised above the hollow. The plant in Plate XII reveals clearly the ethereal space enveloped by the leaves within which the flowers appear. Below, pressing down into the dark earth, is the very different form of the root. At the point between root and shoot, as in the case of the Plantain in Plate V and of the Beetroot in Fig. 14, the plant is drawn together as if into a primal node (the hypocotyl). Below this the roots radiate downward, tough and substantial, with a hard core – the most material part of the growing plant. More often than not, the Dandelion reveals its ethereal space in a conical gesture.

In Plate VI the cone-space is revealed in another aspect; instead of discrete cones following one-another between the vertical infinitude within and the horizontal plane, we see a whole surface – a spiralling cone – which, if pictured in its entirety, opens out continuously from the vertical axis to the horizontal plane.[56] Many plants unfold their young shoots with the spiralling gesture of an unfolding sheath; it is typical of many monocotyledons. In Plate VII, the Canna (*Canna indica*) is drawn to illustrate this kind of unfoldment; the two plates should be seen in conjunction with one another.

Chapter VI

THE STAFF OF MERCURY

42   *Root and Shoot — Radial and Peripheral Formations*

Having gained a certain insight into the type of form belonging to the more ethereal, leaf-bearing regions of the plant, we come now to the polarity of root and shoot. In the root too, in its finer organs, the plant lives and grows by virtue of ethereal, suctional forces, but the whole orientation of the latter is different. Adapting themselves to the earthly-physical domain, roots bring about predominantly radial, densely filled forms, adapted to physical space. Radial forms are the natural expression of the material and physically spatial world, where both the forces and the movements go along lines from point to point. We see the same in human works of engineering, which, being well-adapted to purely physical ends, represent Nature's physical aspect as it were in quintessence. Tube and pipeline – crank, girder and chain, "strut and tie" – these works abound in radial forms, mechanical and kinematical.

So too the living body produces radial organs where it enters into strong relationship with earthly forces. From the most rudimentary pseudopod of the amoeba to the highly organized limbs of the vertebrates and man, organs are formed to push and pull, to produce and to withstand the mechanical forces of the inorganic world. In the plant the radial quality reaches upward from the roots into the higher organs, both to sustain their weight and to transmit the fluids to and fro. Stem and branch, petiole and veins of leaves are wisely adapted to their mechanical function, where the sheer weight and leverage of spreading forms have to be borne and transmitted to the central stem. Indeed the pure line and structure of the mechanical forces are often more evident in the shoot than in the massive, gnarled or tangled root, but this is so in human architecture too, the farther upward we lift heavy matter from its normal home. A cantilever bridge, an arch, a girdered roof make manifest the play of forces precisely because the structure has to be reared into light and air with a minimum of material – into a realm where such material of its own accord would not rest.

Such then is the polarity of root and shoot. Below, the living form adapts itself to the character of physical space and earthly forces. Above, while even the most delicate leaves and petals still have to be borne aloft by radial and earthly organs, yet they immediately adapt themselves to another form of space, ethereal and peripheral. It is an interlocked polarity such as Goethe recognized for light and darkness in the realm of colour, each of the polar opposites diving into the realm of the other and manifesting from within it, as well as in its own domain.

The living interplay of radial and peripheral formative processes, related as they are to the character of physical and ethereal spaces respectively, will obviously be of prime importance, and we shall therefore need to study this on a more simple and elementary level before proceeding to the essentially bipolar morphology of the higher plant. We will therefore begin with a more concentric form of life – a single living sphere, such as we find among the radiolaria in the animal kingdom (Figs. 56 and 57). (Or we might also think of a two-dimensional, circular, leaf-like form.)

In the purely geometrical aspect, such a sphere, as we have seen, is formed in a dual way – pointwise and radially from the centre outward, planewise and plastically from the periphery of space inward. We will suppose now that the two components – radial, and peripheral or tangential – are not only dynamically active but are made visible in material organs or tissues differentiated from the rest of the body. The hardened and supporting organs of a living body may very fruitfully be regarded as containing the two morphological components which we will name: radial and spherical, or radial and peripheral. Some, like the long-bones in man, lay more emphasis upon the former; others, like the skull-bones, upon the latter. We have already mentioned the significance of the peripheral formations – skin or carapace or exoskeleton – in relation to the ethereal.

The peripheral will always be in evidence, for the living body is differentiated from its surroundings by some kind of skin or at least surface-layer of biological importance, the plastic form of which will be related to the ethereal forces. The radial will not always be materially manifest – least of all in the delicate beginnings of life – but it will tend to become so when in the course of development the living body comes into closer relation with the conditions of the physical and mechanical world. Nay, this will often happen without apparent purpose, as if by sympathetic mimicry of these conditions. We will suppose henceforth that our ideal living sphere does contain radial formations emanating from or at least oriented towards an internal centre. The radiolaria are good examples.

To give a rest to the rather overworked words, centre and focus, we may here borrow an astronomical term and describe this radiating centre as the "radiant". Two alternatives present themselves. The physical radiant may or may not coincide with the ethereal point-at-infinity. If the concentric, spherical or circular picture in its simplest form is to apply at all, we must assume, however, that it does. For the radiant is not likely to be a mere passive junction; hence with the other alternative we should come back again to a bipolar body and to the morphological type of the higher plant – this we shall follow up a few paragraphs further on.

We will imagine therefore that the centre of the sphere is both the point-at-infinity of the ethereal space of the living organism and the "radiant" from which some kind of physically differentiated forms ray out. In their ethereal aspect, these radii will now be lines-at-infinity, in the sense of paragraph 20. If they are physically hardened, whether as vessel or more solid fibre, we have the paradox that the infinitude – as it were, the ideal void – of the ethereal space coincides with a region of physical fulness and condensation. If they are hollow vessels, it is true, the ethereal lines-at-infinity only run along their axes, not in the actual substance. We are reminded of one of the main polarities of root and shoot – the cylindrical nature of the stem, the central core in the root (paragraph 6). But in either case, whether the matter be within or close around it – the innermost radial line is now an axis of physical strength and condensation. The stem too, after all, is not only a "spiritual staff" of life; it is a physical staff also.

Geometry alone will not tell us what is the real interplay of physical-material and ethereal activities. It can at most interpret and enlighten what we must read, in the last resort, from Nature. Reading from Nature, we declare that the paradox we have just voiced is also among the constantly recurring Ur-phenomena of living form. It is by no means always so; but *Nature very often condenses her physically radial organs in or about the lines-at-infinity of her ethereal spaces.*

(We may observe in passing that endo- and exoskeleton – concepts that apply of course more to the human and animal kingdoms than to the plant – now have this dual aspect: In our ideal living sphere they would be formed, geometrically speaking, by the lines-at-infinity of the two kinds of space: the former radially, pointwise, along the lines-at-infinity of the ethereal space; the latter spherically or peripherally, planewise, as from the infinitely distant lines of the physical Universe in which the creature lives. If the lines in question are of finite number and configuration the outcome will be, for the exoskeleton, some kind of polyhedral form. It would be interesting to study some of the radiolaria, pictured by Haeckel with such care and skill, from this point of view (Figs. 56, 57). For the rest, even the rounded forms of skull and carapace often show traces of an underlying polyhedral pattern.)

Many diversities will arise through the different kinds of interplay of radial structure from within and ethereal formative forces from without. We may imagine these, beginning again with the simple picture of the sphere. These are the obvious variations: The radial structures may extend to the main surface of the living body, or they may penetrate beyond it; or they may be restrained short of the surface, the fibres fading away or bending round until they run more tangentially and even undergo mutual anastomosis. The surface region will in this case be more continuous and smooth, as though the ethereal and plastic forces would here predominate. If the radii reach out more strongly, various forms of indentation will be produced, revealing different kinds of interplay of radial and peripheral components. In the extreme case there will be prickly and spiny forms; the plastic continuity may even seem to vanish altogether. Yet it will still be there, in that the radii do not extend at random. If we imagine a continuous surface drawn through their terminations, a plastic form, characteristic of the living species, will appear, and this indeed we nearly always see. We shall return to this thought shortly, when we contemplate the forms of leaves in the two-dimensionality of the plane (§48).

Returning now to the ideal picture of a living sphere, let us look at the second of the two alternatives, namely, *when the physical radiant does not coincide with the ethereal point-at-infinity*. If the main ethereal focus does *not* coincide with the radiant from which the radial organs proceed, a bipolar body is the outcome. The radiant, being the source from which the radial organs have grown or upon which they converge, will be a focus of life in its own kind.

The bipolar body will thus contain two kinds of centre or focus. One of them will be predominantly ethereal and for this very reason will be of greater importance for the forming of the body as a whole; or as we have seen, the ethereal forces work the more powerfully, the less the focus of their space is physically formed, – the more there is of "chaos" in this region. The other focus is the radiating centre; it too is ethereal in nature, for without this the radiating organs would have no origin for their life and growth. Yet in this region, the physical is more in evidence, though we must remember that in a living body the polarity is always interlaced: the more physical realm has its ethereal and the more ethereal its physical component.

Place the seed – the little point-like entity, seemingly composed only of material substance – into moist earth; give it over to the creative powers of the water, and presently the living forms will reveal the interweaving polarity of physical and ethereal spaces. Water as such tends to form a *sphere*; the spherical drop is like a microcosmic picture of the great water-sphere of the Earth, and as we have seen, the sphere is among the primary forms revealing and evoking the all-prevailing polarity of point and plane. We shall therefore not wonder that this polarity is made manifest in living form when in the moisture of the Earth the seed is brought to germination.

There is much variety in the structure of seeds and so there are differences in their mode of germination, but basic to them all is the form of the young plant within the ripe seed: it consists of a shoot (the plumule) a root (the radicle) and either one or two seed-leaves or cotyledons. In all cases, the same conditions must be fulfilled before germination will take place: there must be moisture, a certain range of temperature, varying with different plants, and there must be free oxygen for breathing. In other words, the seemingly dead little physical form, the seed, must be given over to the totality of the other three elements – water, air and fire – before any living forms may be conjured from it. What then happens will, however, always be the same, though the *manner* of achieving these same ends will show many differences. After the seed has been well soaked for a time, the radicle will appear first and grow down to become a well-developed root, which will later give off side-roots: then, secondly, the plumule will grow up to form the first shoot (Fig. 63).

A detailed study of the different forms of germination reveals in many fascinating ways the development of the bipolar form of the embryo, in which the consolidated form of the root may be seen in contrast to the little plumule,

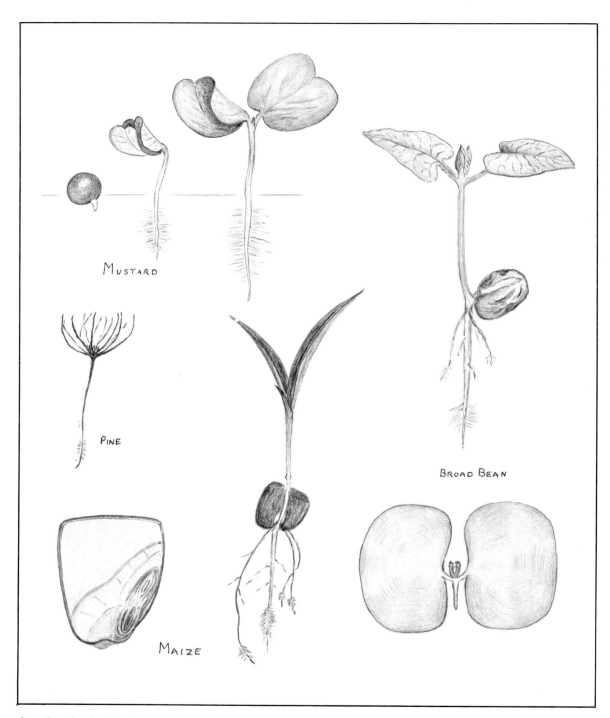

MUSTARD

PINE

BROAD BEAN

MAIZE

63   Germinating Seeds

formed of planar organs, which – sometimes even already within the seed – reveal the hollow-space gesture containing the ethereal focus. Typical on the one hand is the naked, densely filled, radial form of the root, growing downward into the soil, and on the other the sheathed form of the shoot, protected as it is in varying ways by bracts or enveloping seed-leaves (cotyledons) or by true leaves. From the very start, the ethereal space containing the focus of life is often to be seen in one of its most ideal manifestations. The germination of a seed is in fact like the coming-to-birth of the bi-polar form – a sundering of earthly and sun-like centres. It is then followed by a quick expansion into the developing green shoot (Fig. 64).

This first stage of development results in what is called the "hypocotyl", the first and primary node. Unlike the Plantain and the Dandelion, for instance, most plants do not simply raise a flower-bearing stem from a rosette of leaves on the ground, nor are the leaves, as in the Plantain, crowded together at ground level, each with a flower-bud in the axil. Frequent as these forms are, most plants develop node after node with lengthened internodes all up the leaf-bearing stem (Plates VIII and XIII), and only then come to the crowning glory of the blossom. Often a far-reaching variation of the leaf-form accompanies the upward sequence of the nodes (as for instance in Figs. 84 and 85), and we must now begin to ask: What is the significance of the node as such in the bipolar morphology of the plant?

Not only is the form and gesture of the whole plant, with its organs above and below the level of the soil at the hypocotyl, a clear indication of the fundamental polarity at this archetypal node, but a vertical section, for example, through a young seedling, will reveal the same polarity in the detail of the contrasting inner structure of root and shoot (Fig. 65).

Where the shoot is concerned we bear in mind what has been said about the cone-space. The gesture of the entire shoot may sometimes be more conical, sometimes more spherical, the stem may be cylindrical, but the structure is always peripheral and the organs predominantly planar. The leaves and branches spring from the cambium which is an outer layer of the stem; the stem itself may even be hollow.

The sections of Brussels Sprout (*Brassica oleracea gemmifera*) in the right-hand corner of Plate III – the horizontal section of a stem at the point where the leaf leaves it and the vertical section of a sprout which has begun to shoot – show the manner of branching of leaves from the outer layers of the stem. In the Dandelion (centre) and Hogweed (top-left) is shown the polarity of root and shoot, below and above the hypocotyl. The horizontal sections on the lower left of the picture show the contrasting structure of a stem and a root. The drawing of the stem is from a young Elderflower shoot, showing the cambium layer in the periphery, underneath the young green bark, from which any new organs would grow. In the later stages of development, when the stems of the Elderflower become woody, the material within them turns to pith and at last the wood becomes hollow, except at the nodes.

The structure of the root on the other hand is radial. The vascular system is internal, forming a central core from which the lateral roots grow out through the surrounding layers (Fig. 66). The root-tip burrows its way into the soil, its protective cap being constantly worn away and renewed. It is like a vital self-regenerating little sword, piercing its way through the darkness.

The point where the plant emerges from the soil into the light is in all respects a threshold and a crossing-point. In the vascular system there is a peculiar interchange of what is inner and outer; the whole plant is here drawn together into the primary node.

43   Rhythm of the Nodes – Levity in Upward Growth

The form and habit of plants arises from the interplay of the two poles – root and shoot, both of them imbued with the ethereal principle of life and growth, but the former entering into a more intimate relation to the earthy-material and gravitational sphere. Let us first try to picture the type of form that may be expected to arise within the predominantly ethereal region of the shoot, in the ethereal space determined by the sun-centre. The forms we here discern will then be modified by the interplay of the more earthly principle from below.

The young leaves that begin, reaching upward to envelop the sun-centre above the growing-point, tend out and away towards the horizontal. Although the shoot as a whole grows quickly upward, *relative* to the growing-point (or to the ideal "star" above it), the movement of the growing and expanding leaves is downward. The little leaves which today reach pertly up above the growing-point from which they come, will soon, when fully grown, be far below it. It is necessary to understand this dual, upward-springing and at the same time downward-unfolding movement. Consider, for example, *Silphium perfoliatum* (Fig. 11).

The young leaves of *Silphium perfoliatum* (a plant related to the Sunflower, originally from North America) hollow out a plastic form, and surround a more or less spherical space which soon opens out and becomes shallower. The leaf primordia, microscopically small on the actual growing-point of the plant, are hidden deep within the young leaves, which grow out above them. In opposite pairs, like sheltering hands, they hold the innermost region from which the younger leaves spring. As the stem grows, the internodes lengthen and the growing-point is carried upward, the expanding leaves gradually leave the region of this inner space and open out towards the horizontal. They are succeeded by younger generations of leaves, so that the delicately formed hollow space at the tip of the shoot still remains, while the full-grown leaf actually reaches the horizontal plane. In some plants, this may be at the surface of the Earth itself, or it may be in the lower regions of the stem, where upward growth will as a rule have ceased. However this may be, we regard the leaf as an elementary planar organ of the ethereal Earth in the sense in which it was described in paragraph 32. From the aspect of our present study, near the Earth's surface there are many potential planes, one above the other. Every such

64   *Developing Shoot (Hypocotyl)*

65   *Vertical Section of a Plant (after Sachs)*

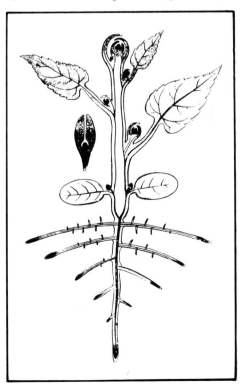

66   *Section of Root Tip (from Wisniewski)*

plane can be a "median" plane or dynamic "plane of levity" for the ethereal space determined by the star-centre above the growing shoot. Think therefore of such a horizontal plane, towards which the growing and unfolding leaves are tending; it will be well below the "star", from which the vertical line of the stem reaches downward, as a rule passing through the plane at right-angles. We picture the star-centre, enfolded by the youngest leaves, the growing-point of the stem a little way below it, and beneath them all the plane towards which the leaves are tending.

Now we imagine this horizontal plane as the common "median plane" of a family of cones, as in Plate IV, or of a concentric family of ethereal spheres in the ethereal space determined by the star-centre, as in Plate IX. The two-dimensional curves in Plate IX serve to illustrate what we will call a family of concentric ethereal spheres. In the plate it is as though we are looking at them in cross section. As we have seen in Chapter III, there is no essential difference between a family of concentric spheres, whose "median plane" is the plane-at-infinity of Euclidean space and the projective transformation of such a family of concentric spheres, where the "median plane" is one of the planes of space. We have taken the top right-hand picture of Plate X as a picture of the former and the curves in Plate IX will serve to illustrate the latter.

As well as the cone space, we now have another geometrical thought form through which to perceive the ethereal nature of the process of unfolding leaves.

The ethereal concentric spheres will not *look* concentric, nor will they look like spheres, judged from the aspect of physical space. Yet they are truly concentric, and truly spherical, in the ethereal space determined by the "star" as point-at-infinity. Seen in cross-section, with the star-centre above and the horizontal plane appearing as a line below, they will look like a family of conics with a common focus and directrix. The "smallest" (that is, ethereally the largest) of them will be nearly circular in appearance, closely enfolding the "star". Thence they enlarge into ellipses, followed by a parabola which opens upward and then a series of hyperbolae flattening into the horizontal line. In the *ethereal* aspect, the latter grow ever *smaller*; they vanish into "zero" when they reach the plane.

Just as we have compared the gesture of the cone forms in Plate IV, which open out from a vertical axis to a horizontal plane, with the unfolding gesture of the leaves of the Plantain, so now we gain another realistic picture for the understanding of the ethereal process of leaf-development at the growing-point, if we imagine the unfolding leaves springing from the central stem between the growing-point and the plane, enveloping or at least touching successive spheroids. In some plants, as we have said, the gesture at the growing-point is more spherical than conical, and this is often to be seen when the leaves at the top of the growing stem are related to a flowering process. In the Woodruff (Fig. 7), for example, it is interesting to see the more spherical gesture of the leaves which are creating the ethereal space for the flower, while the lower nodes take on the conical gesture more typical of foliage leaves up and down the stem of the plant (see also Plate XIII). In the forms sometimes to be seen at the top of the shoot, and very often in rosettes at the ground, each leaf, as it unfolds and grows, envelops ever flatter and more open spheroids, beginning near the star-centre and opening out towards the plane.

Meanwhile the stem is growing upward; fresh leaf-buds are arising at the apex and in their early stage of growth enveloping the physically smaller spheres.

In the early stages of growth the internodes will not yet have lengthened out. The star-centre will be very near the horizontal plane; between the two the spheroids will be crowded very close together. We therefore begin with the same picture on a very small scale and, while expanding it without essential change of shape, simultaneously imagine the unfolding movement from the inmost forms to the larger and flattened ones. Combining the two transformations in this way, we have a picture of what happens as the shoot springs upward and the leaves unfold, down and away from the growing-point. For a beginning, we may imagine the horizontal plane fixed in position. In the resultant, till they have reached the horizontal, all the growing leaves are carried upward; the upward-shooting of the growing-point more than makes up for their downward unfolding from sphere to sphere.

We here approach the kind of interplay between two opposite processes which was described in paragraph 5. This is indeed a characteristic of the life of plants to which our eyes will be opened ever more and more. It is the interplay between two activities – both of them physically and ethereally determined but in different ways – which both in quality and gesture and in their quantitative outcome partly tend to cancel one another out. It is so above all in the vertical dimension of unfolding growth.

Trying to think and feel in the qualities of ethereal space, let us now consider what the two components of this double process mean. The passing of the growing leaves from sphere to sphere, away from the "star of life" and flattening towards the plane, is for the ether-space, as we have seen, a spatial lessening. In the ethereal volume they contain, and in ethereal area, the hyperbolic spheroids grow ever smaller as they flatten; indeed they reduce to "nothing" when they melt into the plane. The more rounded spheroids (or the tiny cones) on the other hand, close to the point-at-infinity, grow ever greater and vaster as we go inward. Where there is physical increase, there is ethereal decrease. The leaves are born in the realm of the ethereal infinitude and thence grow out and downward, until they alight, so to speak, in the plane of Earth. It is as though the microcosmic Sun-space of the star were giving them to the Earth. As they grow larger physically, the ethereal space which they envelop grows ever less. This mathematical and quantitative aspect answers to what we see and feel: the young leaves with their tender yellow-green are replete with life; the fully opened ones with their darkening colour do indeed show the characteristic form of the plant more fully, but they have grown less vital – they are becoming ever more like finished, ultimately dying pictures of the archetypal form.

Thus, we reach the conclusion that whereas the upward growth of the stem, increasing the distance from the star-centre to the horizontal plane, appears in physical space like a pure expansion, what it means for the ethereal space determined by the star is none other than a primary phenomenon of levity or "Leichte".

The relative Earth-plane to which the leaves are tending is an important plane of levity in the individual organism of the plant. The celestial sphere on the other hand functions as a universal plane of levity: for the levitational fields of force it is in polar analogy to what the Earth-centre is for the terrestrial phenomena of gravity. The plant's individual plane is drawn out towards the cosmic plane. This finds expression in the upward movement of the star-centre. Ethereal, the upward growth brings the horizontal leaf-planes nearer to the heavens. The Earth-plane itself is in spring and summer breathing ethereally outward by virtue of the plants that grow upon it. Once we perceive the true idea, we see that it does so qualitatively even where it stays seemingly at rest.

For the spheroids – and their enveloping leaves – between the horizontal plane and the star-centre, this functional outward breathing of the Earth-plane involves an actual upward movement. Even by this rudimentary and no doubt over-simplified thought-picture we have interpreted the growth of plants with a quality of thought far more intimately related to the phenomena than any atomistic thought-form.

To sum up these thoughts concerning the upward growth of plants: raised from the region of convex growth where the plant is rooted in the earth, the physical forms of the leaves are so adapted as to envelop the ethereal spheres or cones in the region adjoining the main stem, between the star-centre and the horizontal plane below. The latter may be any horizontal plane above or at the surface of the soil. For the ethereal space of the star-centre, it is the plane of levity towards which the leaves are tending. But the plant with its ethereal spaces – convex and concave, root and shoot – is itself an organ in the larger cosmic organism of Earth and Sun, and as in any complex living body (cf. paragraph 33) the individual formative spaces, reaching the zenith of their activity from time to time, are so co-ordinated as to contribute to the larger whole, so the leaves, born in the region of the plant's own sun-like space, are given over as they mature to the planet as a whole in its organic relation to the Sun and the surrounding heavens. They play their part as planar organs in the Earth's levitational field (paragraph 31); the shoot as a whole is drawn up and outwards towards the cosmic plane of levity, the summer sky.

We take hold of a truly cosmic imagination of vital significance for the understanding of plant development if, while watching the visible upward development of the plant, as it grows taller in physical space, we realize that *this is only one aspect of a twofold process*. The leaves appearing at the top of the growing stem are, as it were, flowing downwards towards the Earth from an ethereal space; born out of the cosmos, they descend to Earth, there to grow larger, more filled with material substance and finally to die away again and disappear from earthly sight. It is a truth which applies to all life as it comes into being in the lap of Mother Earth, spends its allotted span of time there and passes away again. Scientists will always be at a loss to explain the appearance of life on the planet, until they take the long stride of thought here involved – a stride which will, however, appear less long as scientific thought emerges from the comparatively short, but gigantically forceful period of nineteenth-century materialism in human history.

44   Goethe's "Spiritual Staff" – "Verticon" and Horizon

Physically, then, the plant is centred in its roots below the surface of the soil; a firm physical structure, they hold it to the earth. Above, enveloped by the budding leaves, is the *point-at-infinity of the ethereal space*, imbuing the young organs with vital forces and drawing them forth to living unfoldment. It is as essential for the sprouting, developing activities of the plant as are the physical constituents of its body.

The higher plant as it germinates reveals at once the polarity of root and shoot with the hypocotyl between. True leaves quickly follow the primitive seed-leaves or cotyledons (cf. paragraph 48) and each time this happens a node is formed. Sometimes the nodes remain close together as in Dandelion, Plantain and plants like the common wild Daisy (*Bellis perennis*); more often the nodes are at intervals up the stem. Biennial plants – Foxglove, for example (*Digitalis purpurea*) or Mullein (*Verbascum thapsus*) – have their nodes close together during the first year, the leaves making a wonderful hollow from which the flower-bearing stem springs with elongated internodes in the second year. The leaves may appear in alternate pairs, or they may spiral singly up the stem. Whatever the phyllotaxis, the principle is the same: the leaf as a more or less plane-like organ, with a bud or "eye" – dormant at first – hidden in the axil between leaf and stem.

In Plate XIII the Pepperwort (*Lepidium campestre*) and in Plate VIII the Dock (*Rumex obtusifolius*) reveal the upward striving of the plant, node after node. Every time the fresh young leaves appear they tend with their gentle cupping gesture the "star of life" which begets them. They do not hold it as their own; their time is allotted them, and they make way for the next generation. Opening out in succession towards the horizontal plane, they become mature. Born of the ethereal source, receiving vital strength from thence, leaf after leaf gives of its life to the earth, remaining behind as the plant grows upward to bedeck the earth with plane upon plane of green.

As the plant grows, it is as though the quality and function of the hypocotyl were drawn higher and higher above the earth. Each time a new node appears, it springs from the one beneath it, and the polarity of root and shoot is in a sense renewed. Even the sprouting of actual roots at the node is a very frequent phenomenon, as for example in the labiates and in all plants from which cuttings will flourish. Goethe describes the upward growth as a gradual enhancement, the cruder substances giving way to finer ones.[57] In rhythmic interplay between the "dark" pole and the "light" the plant grows, at each stage raising earth-matter to become more and more receptive of the light.

As Goethe sees it, the plant organs are ranged around a "spiritual staff". How true this is! He speaks of the "law-giving power in the midst".[58] The central axis of the vegetative part of the plant – a region so often hollow or only containing pith – is the "infinitude within" of an ethereal space. This is not a physical or material staff. It is the line followed by the "star of life" as it draws the plant-forms forth; along it are the nodes with their potential powers of development. This is the "Staff of Mercury" – the "Caduceus".[59] The old symbol stands for the forces of healing of the green leaves as they arise in rhythmic sequence, spiralling up the stem of the plant.

The Dock in Plate VIII revealing the staff of Mercury, shows also how the bud in each node, *like a seed in the soil*, may be called forth by its own "star of life" to develop staff upon staff of living forms.

In physical space, the two-dimensional analogue of concentric spheres will be concentric circles in a single plane, and as the former, growing infinitely large, tend to the plane-at-infinity, so do the latter tend to the line-at-infinity of the plane. In ethereal space, the two-dimensional analogue of concentric spheres will be concentric cones (Plate IV, Fig. 41). These will grow infinite *inward* instead of outward, tending in fact towards a focal line – the line-at-infinity of the point which contains them. It is the common line of the said point with the point-at-infinity of the ethereal space as a whole (cf. paragraph 34). The "line-at-infinity within" is the ideal axis of the stem – Goethe's "spiritual staff". It joins the node from which the leaves are springing to the star-centre, poised in the open space a little way above it.

With respect to the physical and ethereal spaces of the Earth-planet, the plant itself, and the locality where it is growing, the vertical line of the stem is polar to the infinite horizon. The celestial plane, determining physical space, contains many lines-at-infinity, and among these some are of special import – for example the celestial equator, the ecliptic, the horizon. For a particular locality on Earth, the celestial horizon is all-important. Likewise in the ethereal space of the shoot, every line through the star-centre counts as a line-at-infinity, and among these one is of special import, namely the vertical line joining the star-centre to the focus of convex growth beneath it and to the centre of the Earth. Recognizing its ideal value in this sense, we can appreciate how deeply Goethe was impressed by the significance of this innermost line of the stem – the "spiritual staff", the "vertical principle", the "law-giving power in the midst".

Just as, for our human experience, there is in physical space one line-at-infinity of supreme importance, the infinite horizon where the Earth-plane upon which we live meets the celestial plane-at-infinity, so for the plant there is one line-at-infinity of supreme importance – the line that joins the star or point-at-infinity to the Earth's centre, or to the point at the Earth's surface where the plant is rooted and whence it germinated. Such is the vertical line of the stem, – Goethe's "spiritual staff". It is the plant's "infinitude within" in its linewise aspect – ethereally one dimension lower though physically one dimension higher than the seed or star. Even as the "infinitude without", determining our own spatial life and consciousness, has the two aspects – two-dimensional, spherical or planar when we look up and out into the star-strewn heavens; one-dimensional and linear when with our consciousness turned to more earthly ends we scan the wide horizon – so for the plant there are these two infinitudes: the star and the spiritual staff, the point and the line, of which the former is primary and contains the latter, just as the heavens contain the infinite horizon.

For the ethereal space that forms the plant, the spiritual staff – the central line of the stem – is therefore an "infinite line within", in polar analogy to the horizon. This helps us understand the vast importance of an axis which is materially often hollow, or in a region of comparatively unimportant pith. "Pith" is a synonym of concentrated essence, the vital nerve of the thing in question; in many languages the words for pith and marrow are the same. Materially, pith does not seem to deserve this reputation. Ideally, we now perceive the significance of this inmost region. It is a formative infinitude, fulfilling a like function in the ethereal space of the plant as is fulfilled in the forming of physical and earthly space by the great circle of the celestial horizon. Goethe divines this when he writes of the roundness of the stem, which, he declares, if not outwardly round, is at least potentially round from within; in it he sees "the law-giving power of the midst".

Even as the geometry of a circle is largely determined by the line-at-infinity of the plane in which it lies, so is each cone or whorl of leaves, arising at a nodal point along the spiritual staff, well-centred and well-formed in relation to this innermost and now vertical infinitude. The youngest leaves, those at the highest nodes, by their enveloping gesture indicate a cone of steeper angle, near to the spiritual staff; as they grow older the leaf-cones open and flatten towards a plane. The cones that are nearest the infinitude within and that look narrowest to physical imagination, are in fact ethereally the largest; the downward-unfolding growth is an ethereal reduction.

Geometrically, as we have seen (Plate IV and paragraph 20) such a family of cones, if carried by a single point, is polar to a concentric family of circles in a plane, which we may take to be horizontal. The vertical "infinitude within" is polar to the line-at-infinity of the plane; the horizontal plane towards which the cones will tend as they widen out is polar to the common centre of the circles.

Once more, as in paragraph 40, it is helpful to relate the ethereal process to the physical by mentally reversing the flow of time. In the physical world we get a movement of concentric circles when we throw a stone into a pond and watch the waves expanding – towards what would ideally be the infinite horizon. Reversing this, we should have the picture of circles tending inward concentrically from the infinite horizon – detaching themselves from thence, as it were, one by one. They lose themselves into the central point. This is analogous to the ethereal process shown by expanding leaves. They come from the realm of the infinite line within, and, with ethereal lessening of the cone-form they envelop as they grow older, will merge at last into the horizontal plane.

Also for the more spherical unfolding forms described in paragraph 43 we can apply the same thought-process. The "infinitude within" is then a point, namely the star-centre. Again and again, the growth of living forms from their seeds, or from the focal regions of unfolding life, is like an inversion – both in space and time – of what takes place in the physical world when entities disperse into the distance. Often when we look down upon the growing-point of a symmetrically growing shoot, it is as though we were gazing into the vanishing-point of some far perspective. We see what Goethe would have called an "open secret". For the ethereal realm which we are here beholding, this is no vanishing-point; it is a point of *origin*, of creation. So too the "spiritual staff". The "vanishing line" of physical perspective drawings most often represents the infinite horizon; the plant's "infinitude within" is a creating line.

This qualitative concept of the stem, or of the innermost vertical line at its centre, is so important both in itself and in its polar relation to the horizon as to deserve a specific name, conveying its geometrical character and at the same time sharply differentiating it from the concentrating and supporting function of a physical-material staff or axis. We shall therefore describe the innermost vertical line, seen in the aspect of ethereal space, as the "verticon", although the form of word may be open to criticism from a scholarly point of view.

The formative space of the archetypal plant – compare what was said of this in paragraph 11 – is spanned between the verticon and the horizon. The living plant is an entity both physical and ethereal, placed in the spatial universe in such a way that the Earth-plane from which it grows, the plane-at-infinity, the centre of the Earth and its own growing-point – or the star-centre hovering above this – are of prime importance: a pair of planes and a pair of points, with one of either cosmically given and one of either given locally or individually. The common line of the two points is the "verticon", the common line of the two planes the horizon. Each of the lines functions as a line-of-points and as a line-of-planes, the points of the one bearing the planes of the other and vice versa.

Both the vertical and the spiral tendency, in Goethe's sense, have to do with both these axes. In uniform circling-measure the points of the horizon determine the circling sequence of spiral phyllotaxis; the same circling-measure belongs to the planes of the verticon, for the controlling character of which the ideal "roundness" of the stem seemed so important to Goethe. The upward-striving growth on the other hand, the rhythmic sequence of the nodes – for the physical aspect of the plant so obviously belonging to the vertical axis of the stem – in its ethereal aspects is related to the successive *planes* of the horizon. It is the horizon which bears the planar, levitational movement which we have just been considering. In other words *the celestial horizon is for each Earth-locality the main ethereal "line of force", relating the horizontal, potentially leaf-bearing planes of the ethereal spaces of the plants to the celestial plane, and thus determining the field of "levity" which draws the plants up and outward.*

A threefold possibility of movement belongs to the physical-and-ethereal space which is thus spanned between the "verticon and horizon". First is the upward-and-downward dimension – in its physical and pointwise aspect belonging to the vertical line in the midst, in its ethereal and planar aspect to the celestial horizon. Second is the circling dimension, determining the rhythmic regularity of phyllotaxis, and in some instances (e.g. the climbing plants, compare paragraph 11) involving the actual growth-movements of the plant.

Third is the radial, inward and outward dimension, revealed in the lateral expansion of leaf and branch and in the conical or spherical unfolding of the leaves from the near-vertical to the horizontal. This dimension too, which we may see imaginatively as one of in- and outward breathing between the innermost and outermost infinitudes, has its ethereal as well as physical aspect. Physically, the spreading leaves and branches carry the sap out from the region of the central line towards the horizon. Ethereally, every time a young unfolding leaf begins in close proximity to the "star" or to the "verticon" and opens down and outward, an ideal plane is leaving the region of the plant's innermost infinitude and tending to alight in the horizon. The rhythm of contraction and expansion appears again in its dual aspect (paragraph 21).

Our descriptions so far have not included the specific phenomena of phyllotaxis – that is, the angular and numerical relations revealed in the spiral sequence of the leaves and in another form in the floral pattern. We do, however, believe that the dual, physical-and-ethereal space-formation – above all, the mutual and polar relation of "verticon and horizon" – provides the framework, as it were, within which these phenomena may receive their true interpretation. The plant – seemingly small as it is in spatial magnitude – reveals the Earth's partaking in a far wider community of rhythms, forces and movements of the great universe, which are not only physical in character. What was described in paragraph 11 as the "spatial matrix" of plant life appears in its true light when the ethereal as well as the physical aspect of space-formation is perceived.

The plant is indeed only *seemingly* small! There are worlds beyond measure not only in the outer periphery towards which the plant expands visibly, but also in the innermost infinitudes of its living form. The plant, as we begin to perceive, when we contemplate it with open eyes and an inner perceptive judgement, comes into being in the interplay of spatial-counterspatial, macro- and microcosmic infinitudes. Far from being limited only to the outer study, however detailed, of the developing form, we are led to perceive in clear thought a cosmic process taking place before our very eyes.

There is another projective process whereby modern geometry paves the way to a clear understanding of the interplay between the two lines at infinity (within and without) which we have called the "verticon" and the "horizon" – the "spiritual staff" in the plant's microcosmic ether-space and the cosmically given celestial horizon.

When considering the polarity between concentric circles in a plane and cones in a point (Plate IV and paragraph 20), we saw how in the limit the infinitely expanded circle in the plane is polar to the innermost axis of the cone-space. The sphere, as well as calling forth the polarity of point and plane, calls forth also what we call the Line-Line Polarity of archetypal space. In Fig. 67 it is shown how a line passing through a sphere has a polar line at right angles to it outside the sphere. Move the inner line to pass through the centre of the sphere, and it will be seen that its polar line moves to the infinite. These are the two limits, and it is in this way that verticon and horizon are related projectively; as lines of free

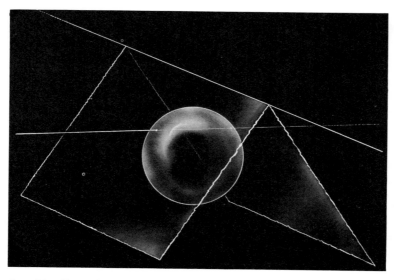

67   *Line-line Polarity with respect to the Sphere*

projective space, they return into themselves somewhat like great circles. Although of infinite extent, they are interlaced with one another in the same sort of way as the interlocked, finite rings in Fig. 68.

Verticon and horizon express the most extreme form of what are called "skew" lines in projective space – *lines which have no point and no plane in common* (cf. the Axioms of Community of Point, Line and Plane in paragraph 18).

Any such pair of lines, however skew and however far apart, will be related projectively, if considered both pointwise and planewise. As a plane of one of the lines turns about it, it will engender a point which runs along the other line, through the infinite and back in either direction. Add to the two skew lines a third, skew to both, and the plane of the first line will meet this third line also, begetting yet a *fourth line*, namely, the line common to the two points – one on each of the other two lines – which are engendered by the moving plane.

As the plane continues to turn, the fourth line thus engendered will move round, sliding, as it were on the three given skew lines (Fig. 69). The movement of this fourth line depicts an infinite manifold of lines, moving right round and leading back at last to the starting-place. This line will plasticize a line-woven hyperboloid surface, opening upward and downward and returning into itself through the infinite. Such a linewise hyperboloid is spanned between two axes, an inner and an outer – those lines to which we have given the name "verticon" and "horizon". In Fig. 70 we have set a family of hyperboloids one within the other, which accords well with their nature. It is revealed that the closer to the vertical axis the form approaches, the less inclined from the vertical are its lines (generators) and the slimmer is the form. Moving out from surface to surface, the generators become more and more inclined, until, if we were able to follow

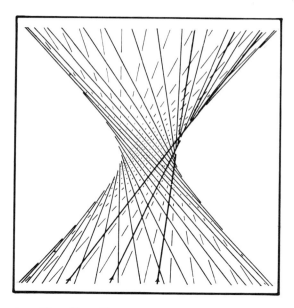

69　*Three Skew Lines*

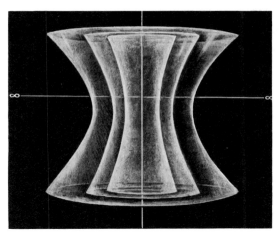

70　*Family of Hyperboloids*

71　*Linewise Spiralling Surfaces*

68　*Interlocking rings*

them right out to the infinite, we should see them all merge into the infinitely distant line at right-angles to the inner axis. It is easy to see how the slim hyperboloids around the vertical axis degenerate into it when they reach their inner infinitude; not so easy to imagine how the outer surfaces fold in upon the outer "celestial" axis, degenerating at last into this other infinitude. In Fig. 71, we see the same process, with emphasis on the spiralling surfaces. This world of line-woven manifolds or reguli, as they are called, is a matrix for all manner of spiralling forms and movements; Goethe was indeed approaching one of the deepest secrets of plant life, when he recognized what he called "the vertical and the spiralling tendencies in plant growth". It is in this realm that thought-forms are given towards the understanding of the cosmic significance of phyllotaxis.[60]

To understand Rudolf Steiner's idea of plant morphology, the counter-spatial nature of the vertical line of the stem – the "line-at-infinity within" of an ethereal space – is essential. In its deeper nature, the unfolding plant, with its green leaves appearing in the sequence of nodes, is not three-dimensional at all; it is so only inasmuch as it is on the way to becoming a material form in Earth space. Just as the celestial horizon perceived at any point on the Earth's surface has an outer, spatial significance for an organism growing at that point, for it determines the *two-dimensional positive-space plane* at a point of which the organism is situated at a given moment, so, for the sun-like space of the growing plant, the line joining its sun-focus with the centre of the Earth is a significant line, for it gives the plant its *negative-dimensional character in ethereal space*. The sequence of unfolding cones at the nodes is a *negative-dimensional* process in ethereal space. The living manifestation of the Mercury staff involves a continual breathing process of expansion and contraction, taking into account positive and also negative space-forms.

We have seen that the essential geometrical thought-processes upon which modern geometry depends are akin to the rhythmic expansions and contractions seen by Goethe in the life of plants. In perspective transformations we alternately expand a form into a plane, where it lies visibly outspread before us, and gather it into a point or eye, where it becomes, in the way we have described it, *intensive* – a form of lines and planes within the point – *invisible* to outer perception, but still perceptible to the activity of thinking.

So too, the plant alternately shows forth its forms and gathers them potentially into the seed or eye; it lives extensively in summer, intensively in winter-time. Moreover in the life and unfoldment of a single plant, the underlying archetypal form is alternately revealed, expanded in the leaf, and gathered up into the potentiality of node and eye. The leaf qua plane is physically two-dimensional, ethereally of no dimensions; the seed or eye, qua point, vice versa. We are reminded again of the "ideal expansion" that goes hand in hand with physical contraction. The stem, bearing leaves and eyes, is the one-dimensional entity both physically and ethereally, joining the two interlacing aspects.

A plastic form, within the physical or Euclidean two-dimensionality of a plane, will be what we call a plane curve – a circle for example or a plane spiral.

Likewise a plastic form in the ethereal two-dimensionality of a point, as we have seen, will be some form of cone. For example, we have shown in Plate VI a very beautiful spiral cone by perspective from the logarithmic spiral, and have been led to the ethereal significance of the cone-forms which are so characteristic a feature in many plants. The points that bear the successive cones, shown by the upward-opening of leaf and branch, are the nodes along the upright stem. Like a plane curve, a cone is, strictly speaking, a one-dimensional form within a two-dimensional space, whereas a sphere or spheroid is two-dimensional within a space of three dimensions.

In the enveloping gesture of young leaves – from among countless examples we mention Woodruff, Lilac, Rhododendron, Dog's Mercury – the plant is using the dual potentialities of space at a less plastic level, just as a plane curve or a cone is less plastic than a true surface. We have also to understand this conical, two- and one-dimensional aspect in relation to the whole ethereal space of the plant with its star-centre above the growth-point. The star itself, as we have seen, is a kind of spiritual seed-point; for the ethereal space which it determines, it is, like the plane-at-infinity of physical space, a two-dimensional form – only intensive instead of extensive. As in physical space every line within the plane-at-infinity is infinitely distant, so in ethereal space every line that passes through the "star" is so (cf. paragraph 34).

This subtle play with the 0- and 1- and 2-dimensional organs of space, both positive and negative, is most characteristic of the higher plant. Precisely this organic dissolution of the plump three-dimensionality of space is rendered possible by the projective interplay of the two polar-opposite types of space. This makes the beauty of the plant, and the many ways in which the plants achieve it contribute to their infinite variety and their effect upon our life of feeling.

The plants reveal this dynamic of the Mercury staff in very different ways. The living, rhythmical process, in which the bi-polar space is continually being renewed at a higher level, where there are very definite nodes, separated by internodes is especially evident in the dicotyledonous plants, such as those we have evidenced as examples. In the monocotyledons, the Mercury staff is revealed in a different way (cf. paragraph 48). The interplay between the forces of light of the star-centre and the physical material of Earth does not reveal itself with such contrast in the Lily as in the Rose. While showing the sequence from node to node far less definitely, the monocotyledons express the interweaving of the two poles in gradual transition from roots to bulb, from bulb to leaves, and then to the flower within the leaves. Sudden contrasts are rare; the poles are far more closely bound together. The succulent, tenacious life of the bulb often gives birth to the rudimentary flower deep down in the earth, long before it appears and becomes fully developed in the light and air. Not the principle of the nodes is pronounced; rather the vertical axis. The monocotyledon bears itself erect and the plant organs are arranged around the axis often with very little differentiation between rootlike organ, leaf, stem and petal. The whole of the very central region of the plant – the "spiritual staff" – seems concerned with the flowering and fruiting process. In *Colchicum* for example the flower extends right down into the bulb, while in Garlic (*Allium vineale*) bulblets are formed high up among the flowers.

Of the two great classes of higher plants it would be true to say that changes of form through metamorphosis are greater in the dicotyledons than in the monocotyledons. In the dicotyledons the two poles – the "light" and the "dark" – are more sundered, with greater opportunity of living interplay between them; in the monocotyledons they are more united.

45   The Lemniscatory Process – Radial and Peripheral Formation

We have now come to the point at which it will be valuable to introduce the idea of yet another geometrical process – one which has, as it were, been at hand all along and which can give wings to our further thoughts concerning the interplay of the two active poles, physical and ethereal, in the bi-polar nature of the plant. We shall be led from the idea of mere *variation* of form to the deeper concept of *metamorphosis*, as approached by Goethe and developed much further by Steiner.

Let us consider the drawings in Plate X. The first two of these have figured very often in our deliberations so far. We will describe them yet again.

The first drawing shows a family of concentric circles, the length of their radii changing in geometrical progression between the centre and the infinite periphery; emphasis is laid on their radial properties. In the second drawing the same circles are shown, but in the polar-opposite way; for every point in the first drawing there is a tangent line in the second. Here the emphasis is on the peripheral aspect.

Picture these drawings three-dimensionally; the circles would then be spheres, pointwise and radial in the one, planewise and peripheral in the other. They illustrate two opposite worlds. In the one the points shoot outward, becoming lost in the infinite distance; or else they concentrate as though attracted for ever inward by a central point, like a centre of gravity. This is the picture characteristic of physical forces; it is typical of physical space.

The second picture illustrates the opposite process. Planes move inward from the infinite distance, spending themselves like waves on a shore as they reach towards an infinitude within; or they withdraw out and away to the periphery. Planes do not move in a manner characteristic of a projectile as points may do. They hover or float inward or outward in relation to the centre. As they come inward they envelop and mould a hollow space from outside, quite opposite in character to the point-filled radial sphere which grows from the centre of the other picture. The outward movement of the planes would have a suctional effect rather than an explosive one. If we think of the points in the first picture as being drawn to the central point as to a centre of gravity, so we conceive a force of "levity" which would draw the planes of the second picture outward towards the periphery (paragraph 31).

We have been concerned to understand the quality of this hollow space; we regard it as picturing an ethereal realm in contrast to a purely physical one. It is a receptive region of a planewise nature; the planes move in unendingly towards an infinitude within, just as in the other drawing the points move out into the infinite expanse. The "infinitude within" is also like a springing source of planar movement as the planes open out and away from it. This centre too is a point, but it is quite different from the matter-filled points of physical space.[61] In terms of physical space there is less than nothing in it. The contrast between the two centres is as darkness to light; if the physical centre is "dark", this one is "light" (paragraph 37).

The interplay between these two active poles, physical and ethereal in the double sense just explained, leads to a fresh archetypal form between them, so that the organism as a whole becomes of threefold, not only twofold nature. It is a purely geometrical form to begin with, but we soon learn to see in it the archetype of an exceedingly manifold and versatile principle of space-formation, perhaps of scarcely less importance in the life of Nature than the sphere- and circle-forms from which we took our start. It is the outcome when spherically formed spaces, ethereal and physical, are interpenetrating between two distinct poles. If radiant and ethereal focus are one and the same point, the formative ether-spheres and the outward-radiating activity are in polar relation to one another, the one tending inward in the same geometrical progression as the other outward. We now preserve this relation when the two poles are distinct, as they are in a bipolar organism. We look for the places of interpenetration of two distinct families of concentrically growing spheres (or circles, in the plane picture). From the physical radiant we let a sphere grow outward — in simple "growth-measure" or geometrical progression. Towards the ethereal focus we let a corresponding sphere *grow inward* in the same growth-measure; *the one is a radial, the other a peripheral process.*

In the third drawing (returning to the two-dimensional aspect) the two families of circles are made to interrelate with one another. Bring the two families into movement in imagination. From the "dark" centre the red circles grow outward, expanding into the infinite periphery; from the infinite expanse the blue circles "grow inward" until they lose themselves in the "light" centre. As they do this the circles interpenetrate and give birth to the forms in the fourth drawing.

To follow this in detail, imagine to begin with a very small red circle growing outward and a very large blue one coming inward. At first they have no real point in common. Soon, however, they will touch in one point at the lower end, nearer in or farther out according to how the outward and inward movements have been timed. From this moment they interpenetrate, crossing in points on either side of the central line, which as they move upward describe one or other of the curves of the fourth drawing. The upper end of each curve represents the moment when the circles again lose touch, the red circle growing outward and away and the blue one disappearing into the "light" centre.

While the red circle is increasing and the blue decreasing a moment is bound to come when they are of equal size. If at this moment they still interpenetrate, they will quite obviously be forming one or other of the oval or indented curves. If they just touch in the middle in a single point, they will be forming the looped or figure-of-eight curve. If the equal-sized circles fail to meet, a double egg-shaped curve inside the loop will be the outcome.

The curves thus formed are the "Curves of Cassini", one among them, the looped curve, being the "Lemniscate of Bernoulli". They are all continuous curves, weaving in the most harmonious way between the two centres, which are their foci.[62]

The Cassini curves are here regarded as expressing the dynamic interplay of two contrasting processes, *radial and peripheral*. The mathematical law of their formation suggests this interpretation, as the pictures show. There is a figure-eight curve placed vertically. The upper and lower loops are equal and at the crossing-point in the middle the curve cuts through itself at right angles. There is a focal point within each loop. Within and around this figure-eight curve and, to begin with, following it closely, the curves of Cassini are drawn – in an unlimited number, one within the other. The curves close outside will be in one continuous sweep, indented in the middle. As we go farther out the indentation gradually disappears; at last we get pure ovals, which, going farther out to the periphery, become more and more circular in form. The curves within the lemniscate are divided into egg-shaped ovals. As we go farther in towards the foci, the twin ovals not only become smaller; they too become more and more nearly circular in outline. We now consider the complete picture – lemniscate and curves of Cassini all together – filling the entire plane. (If we rotate it about the vertical axis we get the corresponding spatial forms; the surfaces close inside the lemniscate will be egg-shaped and the indented ones close outside it like an hour-glass.)

It is important not to remain bound in imagination to the rigid forms but to see each curve, once it is created, as a stage in a process of metamorphosis. Not only do the forms arise as the result of dynamic movement, but they are themselves in movement. Each curve represents a single moment in the change of forms one into another. The curves flow into one another in a sequence of metamorphoses. Unmoving, however, like two stern guardians in the whole process are the two points, the "dark" centre and the "light", the influence from each pervading the whole in varying measure. Below, the curves are more in the realm of the radial activity; above, in the peripheral. Constant also is the infinite circle of the plane, for as the curves move outward they become more and more nearly circular, until they echo away into the infinite periphery of space.

A word must be said about the double egg-shaped curves inside the loops of the lemniscate. Each pair, one in one loop and one in the other, form a single curve which is in reality just as continuous as the lemniscate itself or the curves around it. To follow round the curve just inside the lemniscate would be to pass at first round the red part, then to cross the intervening space and follow round the blue. The underlying mathematical process reveals that this is possible: in the realm of idea the seemingly divided curve is continuous,[63] though in its most external aspect it manifests in two distinct parts, the one oriented more towards the "dark" pole, the other towards the "light".

We began by picturing the first two drawings of this plate three-dimensionally and it is necessary also to do this for the third and fourth drawings. We then come to the idea of a Cassini *space*, expressing the interplay of pointwise and planar, radial and peripheral activities. Seen in this aspect, the Cassini form is indeed an emblem of one of the most essential polarities in Nature. The forms pervade the whole of space, flowing through it like waves, anchored in the two centres and dying away into the infinite sphere around – the infinitely distant plane of space. We learn to create this geometrical picture accurately in imagination. It must not remain static. In bringing the forms into movement we learn to see them as *gesture*. It is not the mere form but this *gesture of form* along with the underlying mathematical idea of its creation which we shall apply to our further contemplation of the higher plants.

This conception may be varied in many ways. For example we can make the potencies or "common ratios" of outward and inward growth unequal instead of equal. A metamorphosed family of curves results; the lemniscate is changed into a figure-eight curve with the one loop larger than the other, the crossing-point being proportionately shifted towards the centre of faster movement, so that the smaller loop surrounds this point (Plate XI upper left). A great variety of "lemniscatory spaces" is thus obtained. We shall now use the names in this wider sense, speaking of "lemniscatory" or "Cassinian space-formations". Needless to say, this refers not to the mere outer forms but to the *dynamic process* by which they come into being. The variety is still more increased if we allow one or both of the families of spheres to grow outward from or in towards an eccentric point (Plate XI upper right). Moreover sundry projective transformations are possible, e.g. the spheres may first be elongated into elliptic forms. It is a realm of "lemniscatory space-formations" arising from the mutual interplay of physical and ethereal spaces – one of the fundamental though not the only type of form to which this interplay will lead.

All that was said of the polarity of root and shoot and of the rhythmicized polarity from node to node will find expression in this form. We draw the lemniscate and its Cassini curves (representing, in a plane picture, the corresponding surfaces of rotation) in such a way as to recall the functional polarity of the two families of spheres. The lower spheres will be predominantly radial and physical; the upper ones peripheral, ethereal. We must allow this difference to be transmitted to the resulting Cassini forms, so that even if (as in the original form of these curves, discovered by the astronomer whose name they bear) the lower half of the picture is outwardly the mirror-image of the upper, the *qualitative* difference is preserved. The lower one of the two foci must be felt as radiating, filling the ovals in this region from the centre outward; the upper one as an "infinitude within", surrounded by enveloping and planar forms. For they are functioning as "radiant" and "sun-centre" respectively. The lemniscatory form with its "Cassini space" will then appear as the natural emblem of the organic unity and polarity of root and shoot.

Let us now picture, in relation to this space, a plant in the early stages of growth, when the first few leaves after the cotyledons have developed but before the stem has shot upward making the internodes distinct. Take the Dandelion plant (Plate XII), or again the Greater Plantain or the Daisy – plants where the internodes are not lengthened between the foliage-leaves. Other examples will be found in most young plants: Lettuce seedlings, Marigolds, Stocks and Wild Mullein come to mind.

The crossing-point of the lemniscate primarily represents the "hypocotyl", where root and shoot meet and the plant normally passes from the dark earth into realms of light and air. Here is the natural interchange of inner and outer as revealed in the anatomical structure of root and stem and in the transition from radial to more enveloping forms (paragraph 42). The young Dandelion leaves are arrayed around the growing-point, their upper surfaces facing inward and often curved as though to mould the hollow sphere. In some plants the invisibly moulded form is more conical; Dandelion and Plantain show individual variations in this respect. In either case we are impressed with the contrast between the wiry, radial quality of the roots and the planewise enveloping gesture of the upper part of the plant, and if we call to mind how the lower half of the Cassini picture was contrasted, qualitatively, with the upper, we see how the whole plant-form – "radial below, enveloping above" – belongs to the "lemniscatory" space-formation.

Ethereal concentric spheres and cones (Plates IX and IV), express the aspect of the shoot in the ethereal space of the star-centre pure and simple. If, as we are now doing, we contemplate the actual relationship of root and shoot, the polar-opposite character of the two main components of plant growth – positive or convex from the root upward, negative or concave from the star-centre downward – will find expression in the lemniscatory formation (Plate X, etc.). This is of course a highly idealized conception applied to a very complex reality. Yet it contains a true ideal key. If a concentric, more or less spherical organic entity is formed by the perpetually living balance of spaces positive and negative, or centric and peripheral tendencies, the separation of two centres will by this same mutual activity give rise to a type of form of which the Cassini space is at least a functional indication; it is indeed often realized to a remarkable extent even in the outer form, as we shall see when we study flower-forms.

The dark and light centres – red and blue respectively in Plate X, green and peach-blossom in Plates XI, XVII and XVIII – are each of them to be conceived in the polar aspect. The centre of convex growth – to begin with, in the root – is not exclusively physical, since in that case it would not be imbued with life, but here the physical is predominating; we therefore refer to it briefly as the physical or "dark" focus and picture it in terms of the radial circles. The "light" or star-centre – focus of concave growth – is to begin with purely ethereal; only at a later stage, through the flowering and fruiting process, does it become imbued with matter. Taking the one realm therefore to be predominantly physical and the other ethereal, the simultaneous process, outward in the one family of spheres and inward in the other, is the natural expression of their mutually polar relation. In both instances the process is determined by the relation of the infinite centre within to the infinite periphery of physical space. The form is therefore not self-polar, since there are two centres, both in relation to the one plane-at-infinity. Also the Lemniscate and Cassini forms arise by the *pointwise* interpretation of the respective spheres, although the spheres of the upper family are planar.

The geometrical progression, being functionally a process taking place between two infinitudes – point within and plane without, in this instance – enables the outward and inward movements to be co-ordinated in the simplest and truest way. This remains so even when the common ratios are different as in Plate XI (top left) or when the one point is eccentric as in Plate XI (top right).

Morphologists have often noted – in connection with the logarithmic spiral, for example – that geometrical progression is natural to living growth. The concept of ethereal space helps us to understand the ideal reason. Concentric growth towards the infinite periphery is organically a process taking place between *two* infinitudes – that of the living entity's ethereal space in the centre as well as the infinitude of physical space in the periphery. Therefore the true mathematical process is here the hyperbolic or multiplicative one with its two distinct functional infinitudes and not the parabolic or additive with only one.[64] If we combine the potentized growth-measure which is thus determined between centre and periphery with the circling-measure determined by the absolute circle (and by its counterpart, the absolute cone in the ethereal centre), the logarithmic spirals, seen in so many living forms (Figs. 52–55) are the outcome, nor need the growth have followed spiral lines to produce them. The logarithmic spirals of composite flower-heads or leaf-rosettes, though not concentric in all dimensions, are concentric in a plane which, to begin with, is ideally the horizontal plane. They are growth-measure forms between the Euclidean infinitude of the horizon and the ethereal in the innermost – the region of the "verticon" or "spiritual staff".

If in the forming of the typical bipolar space as in Plate X the potentizing rates, or common ratios of geometrical progression for the inward and outward movements are the same, it is mathematically obvious that the Cassini forms, the "curves of constant product" must result. But the essential thing is the geometrical progression as such, and the co-ordination of inward and outward movement in the two families of spheres; hence if the ratios are different a transformation of the same "lemniscatory" type of space results (Plate XI).[62]

The combination of physical outward with ethereal inward movement suggests that while the plant is physically growing it is being endowed, through its star-centre, with ethereal energies from the great Universe. This combination of physical outward growth with ethereal inpouring from cosmic sources seems to accord with the predominantly anabolic physiology of the green plant, which in its growth not only renews its own life but provides nutriment for other creatures. The interrelation of outward and inward processes need not, however, be thought of as taking place in external time; the time-parameter in the description, and in the mathematical deduction in Note 62 is but to indicate the functional correlation. What does take place in time in the ethereal region of the shoot is the unfolding and expansion of the leaves as described in paragraph 41. (It may be mentioned in passing that if in the process indicated by the bottom left-hand picture, Plate XI, the direction of the upper spheres is reversed and both families of spheres tend simultaneously outward, the resulting forms are reminiscent of the invaginations of animal embryology.)

Let us turn once more to the rhythm of the nodes. For the development of the shoot, the hypocotyl is a first and primary node. Every node shares something of its nature, as a "nodal point" in the prevailing physical-and-ethereal polarity. (Paradoxically, as often happens with the first member of a homologous series, the hypocotyl as the initial node is archetypal and yet in obvious respects atypical.) Consider how each node evolves in the growing plant. It begins its existence in an almost microscopic region where at the growing-point of the stem the primordia of the leaf-buds are born. It is at this moment beneath the sun-centre and above the more physical and radiating realms of root and lower stem which carry it and provide substance. Ideally it is poised radiant below and the sun-centre above, like the lemniscatory crossing-point in the bipolar space. But at a later stage the node becomes more physical; it becomes part of the supporting ground by which the younger nodes are sustained. From being poised as it were between Sun and Earth, it now condenses to a more earthly form and function; so it becomes a radiant in its turn and indeed appears as such, more or less markedly according to the form of plant.

Now the radiant, as we have said before, is at once a focus of life – life in which the physical predominates. Therefore the plant has power to generate ever new centres of such life as it proceeds from node to node. It derives this power from the relation of its earthly nature to its "sun" or "star", which – for the single shoot – remains one and constant. Moreover, in the ideal bipolar space, it is at a lemniscate crossing-point that each new focus of earthly life is born. The crossing-point is in fact a new potential "point of chaos" which then becomes a centre of physically radiating forms – in other words, a living radiant. This now enables the lemniscatory space to be potentized, renewed again and again at a higher level – from node to node up the stem.

(The potentiality we here attribute to the lemniscatory crossing-point is suggested also by its mathematical properties. In the Cassini-space as a whole it is a singular point with a certain quality of chaos. A differential invariant becomes zero as we approach this point; the differential coefficient becomes indeterminate when sought by ordinary methods. The mathematical properties of this region are in harmony with its organic function.)

We gain a beautiful and very suggestive geometrical picture of these relations if, as in Fig. 72, we begin by drawing a large and simple lemniscate of Bernoulli, taking this curve to typify the whole Cassini-family of which it is a member; the same will presently apply to its successors. We indicate the radiant below, the sun-centre above, bringing out this qualitative difference in the way we depict the focal region within either loop. Now we recall that the crossing-point is to

72   *Caduceus Lemniscate*

73   *Decussate Form of Phyllotaxis*

74   *Stinging Nettle*

75 *Botrydium granulatum. Fresh-water alga*

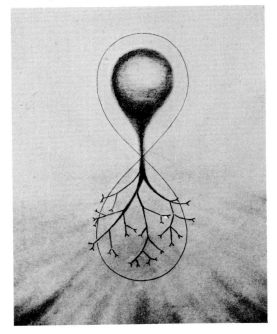

become a fresh radiant; we therefore draw a second lemniscate, with the top focus – that is, the sun-centre – unchanged but with this crossing-point for its lower focus. The new lemniscate will of course be half as large; its upper loop will be entirely within that of the first.

Now we repeat the process, using the new crossing-point as a new lower focus. We obtain curve after curve, one within the other enveloping the same sun-centre, and each one with its earthly radiant raised to the level of the previous curve's crossing-point. If in the quality of the picture we have expressed the underlying thought, it will tell its story. It is of course, like all geometrical emblems of this kind, too static, for (as in paragraph 43) we should have to imagine the whole picture, or at least the upper part of it, simultaneously growing and expanding as we pass up and inward from node to node.

Revolved about its central axis, this family of curves reveals a formative principle which is at work in many variations in the plant world. Seeing each lemniscate in the polar aspect we have explained – radial below, enveloping above – it shows the inner meaning of the "cup within cup" formation described in paragraph 8. Moreover the whole picture, reminiscent as it is of the "Mercury staff" or "Caduceus" form traditionally associated with the leafy region of the plant, gives to this ancient symbol a very significant, functional meaning. If we now rotate the successive lemniscates about the vertical axis, each through a constant angle from the last, we shall be able to follow the various number-relations and types of phyllotaxis; the planes will indicate the main directions of the leaves – or rather of their petioles – growing in spiral sequence from the nodes. Fig. 73 represents the decussate form of phyllotaxis, to be found, for instance, in the Stinging Nettle (*Urtica dioica*) in Fig. 74.

Node after node, the plant raises the dual potentiality of the hypocotyl higher above the surface of the Earth. Each new node gives the basis for further vegetative development; every side-shoot, growing in the axil of a leaf, springs from the node even as the main shoot springs from the hypocotyl (Plate VIII). The node does indeed contain germinating powers, as is well-known from the practical use we make of this in propagation by cuttings. Spanned in a rhythmical interplay between its own sun-centre and the earthly realm from which it grows, the plant thus rears its living form and substance from the ground. This is the process Goethe sees when he describes the upward growth as a gradual enhancement, the cruder substances giving way to finer ones until the greatest refinement is achieved in the flower.

This "Caduceus" sequence of lemniscatory spaces from node to node is an interesting and natural further development of the basic idea. The lemniscatory node or double-point is a unique point of balance between the primary foci in the Cassini space as represented in Plate X or in the other pictures. This very point, we suggest, tends to become a fresh ethereal focus within the living body, in other words a fresh centre of "convex growth". In such a centre, acting as the "infinitude within" of an ethereal space, there must be something of the quality of void and chaos (paragraph 40). Now in the mathematical form of the Cassini space the region of the lemniscatory node has special properties. The neighbouring curves, save for the one which goes exactly through the point, tend to avoid this region. Moreover if by differentiation one seeks to ascertain the slope of the two curve-branches passing through the node, one gets an indeterminate value, due to the fact that at this point a certain differential invariant becomes zero. All these things indicate that here may be a realm of new beginning. The mathematical and botanical conceptions of a "node" coincide; and experience confirms what the mathematical idea suggests – the germinating power of the plant is renewed. The entire plant is, as it were, newly planted. The lemniscatory process, working originally between the star-centre and the root as indicated in Plate XII is thus repeated between the star-centre and each successive node. The star-centre as such is not repeated but remains as an inexhaustible fount until at last the flowering and fruiting process receives it into the body of the plant.

It is in the very nature of the higher plant that in the rhythmic repetition from node to node or in some other way, the plant *takes its time* as it goes towards the culmination of its life in the flower and fruiting process. Herein lies the possibility of metamorphosis. There are lower forms of plant life, the fungi, for example, which do not do this, and hurry towards fruition. Fig. 75 shows a tiny fresh-water alga – the almost microscopic *Botrydium granulatum*, which grows on moist clay in dried-up pools or swamps. Within its single cell-wall it imitates the higher plant's polarity: downward a branching system of root-like organs, upward a single, green, pear-shaped sphere, within which spores are formed. It is again a "lemniscatory" space-form – radial below, enveloping above. But the focus of the upper sphere is not left in open air; it is claimed at once, drowned as it were in the watery realm to which the plant belongs. The comparison of this with the typical form of the higher plant would be achieved by slicing off the upper portion of the sphere and re-forming the resulting chalice to a ring of leaves or to a kind of calyx. The "star of life" would be restored to realms of light; instead of teeming spores a flower would be formed to prepare fruit and seed in more stately measure.

Such transformation cannot be made in a single stage; rhythmic unfolding of the leaves and nodes precedes the flower, taking its course in realms of air and light during a certain period of time. This is the rhythmic process indicated in the "Caduceus" – the potentized lemniscate-formation – and we can well understand the traditional relation of this principle in the higher plant to the powers of healing, pictured in Mercury's staff. The higher plant attains its virtues by tarrying in the realms of light and air, taking a longer time to unfold and to mature through all the stages of metamorphosis.

46   The Hidden Vortex

There is much in the plant shoot which is reminiscent of spiral- and vortex-formation. If we follow the arrangement of leaves up the stem, we move upward in a spiral (even if, as in the decussate arrangement, we go round the stem at ninety degrees a time). But this actual movement as such is not present in the plant, nor even in the movement of the sap (except in microscopically small forms), for in the main the cells run vertically up and down the stem. At most, the movement of the sap in twining plants follows an outward, spiralling path. But twining plants are rather apart; they are only possible, because other plants in which the "law-giving power in the midst" does rule, provide them with vertical stems and trunks around which they may twine.

There are times in which even the higher plant — and a free-standing one — loses its virtue, that is to say, loses its beautiful and healthy balance between the vertical and spiralling principles of its life. This becomes manifest in abnormal growths, and it is in the abnormal forms which may appear in plants that the plant flaunts her secrets abroad for those to see, who, like Goethe, have eyes to see. Forms of growth may arise, which reveal that even in the normal plant forces are latent — brought to rest by the "law-giving power" in normal circumstances. Fig. 76 shows a Valerian plant, discovered and pictured by Goethe. In this swollen stem, the sap surely does flow in a spiral. Such fasciated stems may be found particularly in wet seasons. It is often possible to find, for example, a Dandelion (*Taraxacum officinale*) or a Common Hawkbit (*Leontodon hispidus*) or other plants in which several flower-stems have grown together to form a thick spiral, at the top of which may be gathered a number of flowerheads. One may often discern a spiralling twist in the great trunk of a tree, which would normally be straight. Fig. 77 shows what happened to a Garden Mint!

The spiralling or horizontal tendency has more to do with the forming of substance than the vertical tendency, which reveals the upward striving of the plant in the rhythm of the nodes to more rarefied forms and more advanced conditions of its life. In these fasciated stems it is as though the vertical principle — its mathematically musical, though silent spiral — has been overcome, and has been drawn down into crude material substance. For the plant the result is that there is no longer the power to create fertile flowers and seeds. The plant's invisible, cosmic vortex, open to the starry worlds, has become cluttered with substance. It has become "blind". In common parlance we even call a seedless flower "blind".

Ideally, a vortex unites and holds in balance the power of the vertical infinitude within, and that of the celestial horizon. In water and air, where there is a vortex, there is tremendous actual movement around the hollow space; in a tornado, there is what is called the "eye" of the storm. Rudolf Steiner has described the plants as sense-organs of the Earth; he has said the same of the Earth's springs. We may regard a vortex as a sense-organ, receptive of cosmic influences (Fig. 78); so long as the inner space is clear and open to the universe, the sense-organ can perceive.[13] *In the hidden vortices of plant life, planetary and starry forces are perceived by the Earth and received by Mother Nature for her children.*

76   *Valerian (from Goethe)*

77   *Fasciated Garden Mint*

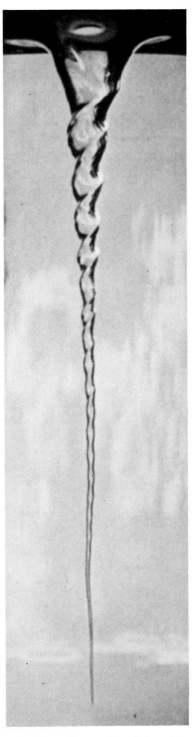

78   *Water Vortex (from Schwenk)*

So far we have been thinking mostly of a single point-at-infinity of the first main stem. In the development of most plants there will be many ether-spaces, as the intensive spaces of the buds in the axils of the leaves open up and form side-shoots. In many plants this development is held back, until the apical stem has reached a certain culmination, perhaps in a flower. Each side-shoot has its ether-space, with the infinitude within at the growing-point, and their formative influences are mutually balanced in an organic hierarchy of power. Every such bud or eye is the living focus for a formative activity from the periphery inward. Again, we must recognize the active interplay between radiant and ethereal focus, and since the ethereal foci are moulding the living forms round and about themselves, the radiant process, reaching up and outward, will tend to conform to their influence. The ramification of side-shoots in plants or of twigs and branches in the case of trees and shrubs will not take place in a haphazard way. Each plant has its own typical plastic outline, and this will not be a mere passive result of the centric forces only, but the outcome of an active interplay; the peripheral forces are always there to meet and "inform" the radial, outward growth. This is precisely the impression we receive from many leaf-forms, from the forms of trees and from the forms of many an inflorescence, such as the flowers of the Umbellate family — the umbel of Archangelica (*Archangelica officinalis*), for example — or of a Composite, such as the Milfoil (*Achillea millefolium*). Here we behold the almost magical continuity of delicately moulded surface made by the many florets, each of them borne by its own stalk to the very place where it needs to be. The polar, physical-and-ethereal interpretation, with the ethereal, peripheral forces directly related to the plastic, planar surface-forms is from the outset far nearer to the phenomenon we see than any merely physical attempts at explanation.

Envisaging the plastic form of the whole plant, if in imagination for a moment we think of every sun-centre shining forth like a visible star, then the more complex plant — with its main outline, its strong branching forms, and its young shoots reaching up towards their "stars of life" — is like the picture of a tree with many candles, symbol of renewed life on Earth.

Most seedlings show the upward-opening hollow cone clearly in the first stages of their development. It is the typical picture of plant substance formed to receive the light-source of its life. In shrubs and trees, as growth continues, the upward-opening conical gesture of leaf and branch is repeated again and again; the forces of the light are received into the living, developing substance. Meanwhile, the tree as a whole gradually assumes its characteristic shape — a downward-opening cone, dark and opaque against the sky (Figs. 79 and 80). Diagrammatically expressed, this twofold process may be symbolized in the picture of two interlacing triangles (Fig. 13). Changed from a dead form into a functional, dynamic image, the ancient symbol of Solomon's Seal relates to this aspect of plant growth. The downward-opening triangle pictures dark matter being formed and moulded by the light; the upward-opening triangle pictures

79   *Outline of Lime in Winter*

80   *Outline of Larch in Winter*

81   Synthesis of Concave and Convex Growth – Acanthus Bracts

the cup of light with the darkness rising to receive and enfold it. Change this straight-lined and angular form into forms of curved outline and the picture will be capable of a much wider application.

In the centres of vital growth, which multiply as the plant matures, the gesture, as we have so often seen, is a hollow cone, or cup – a *concave* form. But as the living substance accumulates, the plant as a whole will form its own individual plastic outline and this will tend to be *convex*. Concave is the gesture of unfolding growth, convex the plastic outline of the resulting form. A beautiful example is the cushion of an Alpine Saxifrage.

The synthesis of concave and convex growth in the higher plant is typical and fundamental. At its young and vital, ethereally active centres, the plant makes concave forms; these hollow spaces it then gives away on all sides as it unfolds, but the living substance thus derived, as it assumes completed form, is moulded into convex shape and substantial form, which is more natural to the material world (Fig. 81). Not only do convex and concave relate essentially to root and shoot, but the entire plant is a synthesis of these polarities, which weave through and through it and into one another.

Chapter VII

LEAF, THE MASTER MASON

48   The Hidden Proteus

"It dawned upon me that in the organ we are wont to call a leaf the Proteus lies hidden, who can reveal or veil his presence in all the forms the plant brings forth. Forward and backward, the plant is ever leaf, inseparably united with the germ of the future plant, so that the one is unthinkable without the other."[65]

Goethe's words contain the key to his morphological studies of the plant. A single organ – the "archetypal leaf" – appears again and again in varied forms; *the plant from beginning to end is leaf, and the leaf is in essence the entire plant*. It is fundamental to Goethe's insight into the morphology of living forms that he constantly sees the whole revealed in the part and the part in the whole.

Regarding the leaf first in the aspect of its two-dimensional lamina, its form and outline is once more the expression of the two opposing principles, centric or radial and peripheral. From the lower regions of the plant the sap rises and spreads into the surface of the leaf in veins, which radiate outward and anastomose towards the periphery. This radiating form, emanating from the mid-rib or, as in palmate leaves, from a definite point, is set over against an outline or contour. The manifold variety in the shapes of leaves is the outcome of the varying interplay of these two principles. We need only add to the picture of the veins that of a peripheral formative process giving the leaf its boundaries. The veins do not spread out indefinitely; it is as though the outward radiating principle met with the resistance of an inward-coming, formative one (Fig. 82). This is most eloquently expressed in leaves in which the anastomosing veins turn inward again at the leaf-margins (Fig. 83).

Leaves are like little finite pictures of an ideal plane, which reaches out to an infinitely distant line! The leaf-blade or lamina is supplied with nourishment from the roots; the sap streams up through the stem and through the branching and anastomosing of the veins, out into the periphery of the leaf. The forms at the leaf-edge give the leaf its individual shape; this is partly determined by the way the veins divide and radiate, and it is clear to see in the forms of many leaves how the veins run out to meet the peripheral force which as it were, rounds off

82   *Leaf Forms – Interplay of Radial and Peripheral*   83   *Anastomosis (Rhododendron leaf)*

the form. In the serrations and hollows at the edges, the veins anastomose most strongly, cease to spread, and give over to the peripheral, moulding force. *The leaf edges are like images in finite space of the infinitely distant line of their plane.*

Thus the physical-ethereal formation of the sun-earth polarity, which we have been studying in the plant as a whole, is made visible again in the leaf-blade. We find expressed in the plane of the leaf itself, the principle of unfolding growth and plastic outline; the whole is revealed again in the part.

It is again an interplay between light and darkness – between the formative forces of the light and the substance producing forces of the Earth. The peripheral, formative process of the light and air, and the centric or radial formative processes of the darker watery-earthy realm, in interplay in the whole plant development, reveal their signature also here in the leaf-blade. Serrations are expressive of the substance-inhibiting activity of the light and air, while the simple leaf expresses the more earthy-watery pole. Simple, unserrated leaves are those which develop their living, sometimes quite tough, green substance in out-spread well-anastomosed surfaces.

The botanist, Gerbert Grohmann, in his book *The Plant*,[52] chose the Ranunculacea, the Buttercup family, as an example of the great variation (he calls it "metamorphosis") in leaf-forms, even in the single plant from the bottom to the top of the stem. We include here two illustrations from his book (Figs. 84 and 85), which illustrate the contrast between the more rounded, lower leaves, with long petioles, and the very differently formed upper leaves, which become more tapering and also more sessile towards the top of the plant. Grohmann describes the expansion in the lamina in the lower leaves, which are most entire, and the contraction in the upper, more tapered leaves below the flower, commenting that the plant does not return to its original form. "It disappears more and more from the world of space."

Describing plant life over the globe from equator to pole, and comparing this with the difference between plants growing low down in valleys and those high up in the light-filled mountains, Grohmann enlarges on this statement.

In the cold, starry, glittering light and air of the high mountains, the height of plants decreases, as is also the case at the poles. The vegetation clings to the rocks and forms prostrate cushions, covered with intensely coloured, starry flowers; the growing season is fast and short. Grohmann paints vivid pictures: "In the alpine landscape, as in the arctic, the flowering forces are exceptional. Spring in the mountains appears as if by magic, bringing Gentians, Primroses, Anemones and many other plants, never to be forgotten by anyone who has had the privilege of witnessing it. Hardly has the snow melted, when we see the little bells of Soldanella nod in the wind. The contrasts of summer and winter, day and night, frost and heat, light and darkness are sudden in high altitudes. Mountaineer and botanist alike share the same experience; they are faced by the rock masses, and with a heightened sensitivity feel the purity and nearness of the firmament." Writing of the arctic, he says: "The masses of flowers shimmer like a magic veil – an exhalation – over the stunted vegetation." Here the cosmic

84   *Scabious (from Grohmann)*   85   *Wall Lettuce (from Grohmann)*

86a   *Variation of Form. Perspective Transformation – Homology*   86b   *Lemniscatory Form about two focal lines*

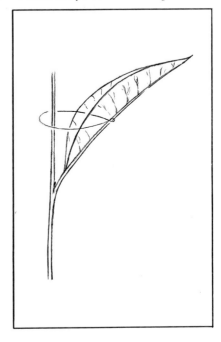

forces are close at hand; the plastic outline of the cushions of so many Alpine plants is a marvel and a joy to behold.

The picture here is of the abrupt meeting of cosmic and earthly forces. The realm of the rhythmic play of the leaves around the Mercury Staff has no place; the rosette or the low cushion is characteristic, the forms unfold close to the ground. "Root-tendency and flower-tendency are juxtaposed but without interpenetration."

The tropics, on the other hand, are a complete contrast to the polar regions. Here the great giants grow; the gigantic trees in the rain-forests are hung and garlanded with a permanent mass of leaves and leafy plants. Here are to be found enormous leaves, as for example, the Banana, and the many clinging, twining lianas. Grohmann quotes a fine passage from Kerner-Hansen, *Pflanzenleben*, which concludes: "High up, the garlands and wreaths of lianas bear the most colourful flowers. A bunch of fiery sparks shows here, a long blue grape cluster hangs there in the sun, and elsewhere a dark wall is embroidered with hundreds of blue, red and yellow flowers; where there are flowers, and where fruits ripen, their guests are never far off – the multi-coloured butterflies and the birds. Their playground is the liana-clad edge of the forest."

In the tropics, the large, fleshy flowers are brilliant, too, but the colours are more earthy; the leaves are tough and of a dark green, sometimes strongly coloured. Here the warm earthy-watery element rises to the tree-tops and provides nourishment for plants, whose soil is dew and rain and rotting leaves.

It is a realm in which substance is dominant, where one great impenetrable mass of leaf, stem, trunk, root and flower intermingles. The flowering impulse works right down into the root, resulting in aromatic, narcotic, even putrid scents, while the roots hang high up in the air. Monocotyledons abound.

Returning to the picture of the Mercury staff or Caduceus, we realize that it is in the temperate climes that the vegetative shoot, the truly balanced plant, thrives; here more plants with a healthy development between the vertical and the spiral tendencies are to be found; many healing herbs with fragrant and medicinal qualities unfold their green shoots in the more harmonious interplay in the temperate zones of the forces of Sun and Earth.

So far we have looked at the leaf as an outcome in all its variety of the radial and peripheral forces working in the green surface of living substance, set out into Earth space, so that the rays of the outer sun may fall upon it. In effect, we have been describing the leaf, once it is there before our eyes. As a visible and manifest form, the leaf, with its petiole and lamina, may thus be accurately described in terms of the morphological approach we have pursued. But fully to describe a foliage leaf, a third and very important part must be added to these two; for the leaf, like the whole plant, is not two- but threefold, and in the usual botanical terminology it is described as such. The leaf consists of lamina, petiole and base, or sheath, at the place where it is attached to the stem at the node. (It matters not that one or even two of these parts may in the event be absent.) A true foliage leaf is recognized by the presence of the bud in the axil; this distinguishes it for the botanist from a sepal or a petal.

Goethe sees the leaf with the bud in the axil as one organ – a leaf is inconceivable without a bud in its axil.[66] We are reminded here of our considerations in paragraphs 20 and 21; an expanded, visible form goes together with one which is intensive, not yet outwardly visible. Just as the whole plant lives out its life in the interplay of extensive and intensive, earthly and cosmic conditions, so also the leaf. One could describe it by saying: the foliage leaf "*subtends*" a bud. They belong together as one organism. When, as we have described it, the bud in the axil then grows up like a complete little plant rooted in the mother plant, it in turn reveals the same typical process (see Plate VIII).

The threefold aspect of the leaf might well be described as lamina, petiole and bud, the petiole playing an intermediary part between the polar organs. At one end it tends the bud, often forming a very pronounced hollow space for this little focus of future life; at the other end it joins the blade and in the majority of leaves continues on as midrib through the surface of the blade, to dwindle and disappear at the tip. The leaf, as we have said, is the ethereal organ *par excellence*; once it is there in visible form, outspread in the light and air, it is the ground upon which new life depends – it harbours the as yet invisible life of the future. This organ, containing within itself the potentiality of "extensive and intensive" – earthly and cosmic – is the "Proteus", the active being, which reveals itself again and again in different forms. It is there in stem and bract, in sheath, sepal, petal and stamen and in all the other plant forms, serving the total plant organism and even the whole plant world, according to need. The leaf is perhaps better described as a good servant, rather than an organism suffering change; it transforms itself according to the nature and requirements of the whole situation, plant nature and environment. It is the Master Builder of the plant world.

We prefer to regard the changing and transitional forms in which the leaf appears up and down the Mercury staff as *variations*, similar in nature to the perspective transformations (Fig. 86a), which change circle into ellipse and so forth, and to reserve the word *Metamorphosis* for the much more radical transformations which occur when the Proteus makes its tremendous leap into the world of the flower. Here the relationship of leaf to inner organs of the flower is at a deeper level, the essence of which rests with the law of polarity; it is the polar reciprocal transformation which can be for us an underlying idea here. We learn to recognize the relatedness of forms which externally would seem to be quite unrelated (paragraphs 16, 19).

49   Leaf-Lemniscate

This pictures the still more fundamental way in which the bipolar aspect of the entire plant is repeated in the leaf, besides the radial and peripheral aspects we have described so far. Leaves show by their gesture the leaf's relationship to the ethereal centre of the plant (Fig. 86b). Their upper and inner surface, especially at the growing-point, is hollowed around the ethereal centre; the midrib stands out distinctly from the back of the leaf. In almost all leaves the veins protrude at the back, while very often on the inner surface a hollow or indentation, very marked in the petiole, continues up the length of the rib. Petiole and midrib with the branching veins are the most material part of the leaf and are concerned both with its firm structure and support and with the transport of nutritional juices upward from the root. The lamina or leaf-blade is of finer form and substance: hollowed upward or outspread to the light of the Sun, it is the characteristic organ concerned in the process of photosynthesis.

Lemniscatory forms are characteristic of the interplay of physical and ethereal spaces and are clearly recognizable in all those features of the plant where watery matter is drawn up and out from root and stem into the leaf-like and expanding organs. We have described the leaves as planar organs and we interpreted their enfolding growth as a phenomenon of ethereal space. If now we look in greater detail, to see how as physical-material forms they fulfil this planar function, we find the lemniscate and Cassini space-form revealed in a new aspect.

We recognized the concave surface of the leaf as part of an enveloping cone or sphere, which if complete would be formed about the spiritual staff of the stem, or, in the younger leaf immediately about the "star" above the growing point. Imagine now a lemniscate (and the associated Cassini forms, notably those that follow close outside it) with the one loop – which we shall take to represent the physical pole in the dual process – very much reduced, as in the top left-hand picture in Plate XI, or as in Plate XVI. Put side by side with this the cross-section of a leaf with strong and single midrib and concave upper surface. The lemniscatory character of the latter will be clearly recognizable, only the larger loop, enveloped by the leaf-blade, is of course incomplete. The smaller loop, represented by the midrib, is very much condensed (Plate XVI).

What kind of surface does this cross-section represent? The lemniscate as a plane curve can be transformed into a plastic surface in either of two ways. Hitherto we have rotated it about its axis, obtaining a three-dimensional form in which the crossing-point remains a point (paragraph 45). If on the other hand we put the original lemniscate-curve, say, in a horizontal plane and move this plane upward, keeping it more or less parallel to itself, the curve will give rise to a surface in which the crossing-point has been drawn out into a line, straight or curved.

Such is the type of surface related to the leaf as the cross-section would suggest, only we must imagine the plane lemniscate changing its shape and size, the proportion of its loops and so on while the surface is being made, so that a more individual and plastic form will be the outcome. It is a looped surface folded figure-eightwise and interpenetrating itself along a line. Towards the lower end where the petiole comes near the stem we must imagine the larger loop growing smaller; towards the upper and outer end the smaller loop gradually lessens and fades away with the diminishing midrib. The leaf-blade only indicates a portion of this surface, which is completed as an ethereal form enveloping the "spiritual staff" and the star-centre. As a matter of pure geometry it should be noted that when the lemniscate and Cassini-curves are made into surfaces of this kind, not only the crossing-point but the two foci become lines. Within either loop there will now be a focal line (Fig. 86b).

It should also be borne in mind that throughout these descriptions, when for the sake of brevity we speak of "lemniscate" or "lemniscatory forms", it is not only the lemniscate itself we have in mind but the associated Cassini forms. One or other of these, and – on the *under-side* of the leaf – notably one of those that follow close *outside* the lemniscate, will be most revealed in the forms we see. Moreover it is not the outer forms as such, wherein of course one might find all kinds of fortuitous likenesses: *it is the process by which the lemniscatory forms come into being*, which is the real test of their true application.

The ethereal space of the star-centre and the spiritual staff now appears in a new light. The leaves as real physical organs envelop it in the same way as the upper loop of the original, vertically placed lemniscate (paragraph 45 and Plate XII), sustained as it is by the lower, more physical loop, while its upper loop envelopes the ethereal focus. The matter-filled supporting line of petiole and midrib is to the hollow region of the spiritual staff what the roots are to the shoot and to the flower. The vertical ideal lemniscate of the entire plant, as described for the Dandelion, reappears in a more lateral and plastic metamorphosis in the form and function of the single leaf. The leaf-blade is the ethereal, the petiole and midrib the more earthly portion – naturally more condensed – and the lemniscate or Cassini-form expresses the organic union of the two. We recognize in this connection many a characteristic feature, such for example as the crescent-moon shape (paragraph 49) which is so often found in the petiole, or the accentuated hollow groove which often marks the midrib-line on the upper surface of the leaf.

This lateral and plastic "leaf-lemniscate" comes into being at the node whence the leaf issues; it is, once more, a lemniscate-surface of another kind, and is united with the upper, ethereal loop of the vertical one described in relation to the Mercury staff, which has its crossing-*point* at the node in question. Petiole and midrib – answering to the contracted loop of the "leaf-lemniscate" – run up the outer surface of this ethereal loop as a continuation of the more radial and earthly element which, from below, bears the leaf upward so that it may fulfil its ethereal and planar function. And from the midrib in turn, the radial element in the branching veins streams out and round into the surface of the leaf. The same lemniscatory function is thus repeated again and again in a lateral dimension and in this way the physical and ethereal, radial and peripheral qualities become deeply and organically interwoven.

The illustrations in Plate XVI, viewed together, picture the leaf in the light of the Cassini space-formation, as we have just described it. In the cross-section of a leaf, taken at right angles to the midrib, the latter is shown protruding below the concave lamina which opens upward. At the bottom of the Plate XVI, Cassini curves are shown, around an inner realm, the lower part of the curves being exceedingly contracted in comparison to the upper.[62] It requires little imagination to see the gesture of these curves revealed in the cross-section of the leaf. The "dark" focus is enclosed by the midrib, while the lamina reaches up as though to embrace the "light" focus. This part of the leaf does not in its physical form complete the upper loop but quite clearly reveals it in gesture.

In the illustration of *Basella alba* on the right, the leaf-blade hollows around the flower-bearing shoot which has sprung from the bud in the axil of the leaf. If this were seen in a cross-section similar to the other example, the flower-bearing stem – or the ideal "spiritual staff" around which it is formed – would appear as a point; seen in terms of the Cassini curves, this point would be somewhere in the region of the upper focus.

The archetypal idea of the leaf is in fact a *lemniscatory surface folded around two linewise foci*, a more physical focal line, the midrib, channel for the material juices, and an ethereal focal line, the line of the flower-bearing stem around which the lamina is extended. This shoot may or may not be physically present, but it is always there potentially in the bud in the axil of the leaf.

It is in the very nature of the leaf that it does not complete this lemniscatory surface physically. The green leaf does not enclose its ethereal focus with its substance, but leaves it open to the light and air, always unfolding away from it.

50   *Leaf and Stem – Sheathing Petioles*

Imagining the hollow cone of a whole whorl of leaves and drawing all their lemniscates together in circular cross-section, we should see the ethereal loops overlapping as parts of a single whole, with the spiritual staff as a focal line in the common centre and the condensed physical loops of the midribs symmetrically arrayed around it and outside (Fig. 91 shows this for three leaves). As the cone opens out, the ethereal loops will grow larger and dissolve away. A numerical enhancement of the fundamental lemniscate-form is suggested: a single loop in the ether-space, subtended, say, by two or three or five physical loops in circular array. (We gain a similar though less symmetrical impression from the cross-section of a single leaf – a rhubarb-leaf for example – with many protruding veins or with a strongly branching midrib.)

Our impression that the young leaf tends an ethereal space within its ventral hollow is made still more realistic when we thus join the physical to the ethereal aspect and see how it comes about. We now understand all the better how it is that every leaf is, as it were, wanting to subtend a side-shoot of its own. In the axil of the leaf the eye is born (Fig. 12). It is most characteristic when this becomes a flower-bearing stalk. There is indeed practically no such thing as a flower-stalk without a leaf or at least a bract to tend it. Even the terminal flower with a leafless stalk springs ultimately from a whorl of leaves. Our concept of the "archetypal leaf" is enriched; we see how true is Goethe's thought that in a sense the whole plant is potentially there when the leaf is given.

The petiole and midrib – physical loop of the leaf-lemniscate – bear the leaf upward from the ground; the leaf-blade as we see it is but the visible portion of an ethereal and more expanded sphere, enclosing the infinitude of an ether-space with a sun-centre at its focus. The leaf originally is like an offspring of the sun-centre at the head of the main growing shoot that bears it. But the sun-centres have a tendency to multiply; the very presence of the leaf begets a new sun-centre, the leaf's own, and from the origin or axil a new "spiritual staff", axis of the potential side-shoot, leads to the new sun-centre. If this side-shoot terminates in a flower, the picture is complete.

A vivid light is thrown upon the structure of the stem itself. Phenomenally the dividing-line between stem and leaf is not absolute; the only absolute about the stem is the "verticon" in its midst, the spiritual staff, which we now see in polar contrast to the physical staff or axis of the leaf's petiole and midrib. Often the stem is not only hollow but even bulges out as in the onion and with its green assimilating surface fulfils leaf-like functions. Let us return for a moment to the cross-section picture of a whorl of leaves. We though of it as the section of a hollow cone, near to the verticon when the leaves are young and widening as they grow older. Imagine now the contrary transition; suppose the leaves grow connate and the cone narrows into a cylinder. The same diagrammatic picture, if we reduce its scale, will represent a hollow stem, ridged like the stems of many plants. The outer cylinder of the stem is leaf-like: that it envelops the region of the pith or central hollow is but an uttermost enhancement of the enveloping gesture which belongs to the typical leaf at its inception.

Consider further. The petiole or leaf-stalk branches out from the main stem. A cross-section immediately below the node would show the stem alone; above, it would show stem and leaf-stalk – two instead of one. (We are assuming for the moment that the eye, in the axil of the leaf, has remained dormant.) Now the Cassini-space, seen in a slightly different aspect, represents not only the mutual relation of two spheres or centres but the severance of one into two, or, vice versa, their reunion. Forms reminiscent of the Cassini curves therefore occur again and again under these conditions and are no doubt familiar. We see them for example in smooth cross-sections of a tree-trunk where the tree was branching.

Returning to the Cassini curves described in paragraph 45, we shall recall that the outermost single oval and the innermost double ones approximated to circles. Now if in following the metamorphosis from relatively small double ovals to very large or even infinite single ones we simultaneously reduce the scale of the picture, bringing the centres nearer together, we obtain a picture of the union of two into one within the limits of a "reasonable" size. We must reduce it to an infinite extent for the two foci to become actually one; at the same time, with due adaptation, what would have been the "infinite circle" becomes a finite and therefore true circle of any desired size. A similar process is seen in Plate XIX.

We get a rough picture of what is meant if we draw a figure-eight curve within a circle, preferably a little smaller than the circle. Fill in the intervening space with oval and indented curves and the space inside the lemniscate with double ovals, culminating in smaller and more circular but still finite curves. The indented curves show the gradual division of the single entity, the lemniscate the actual moment of severance.

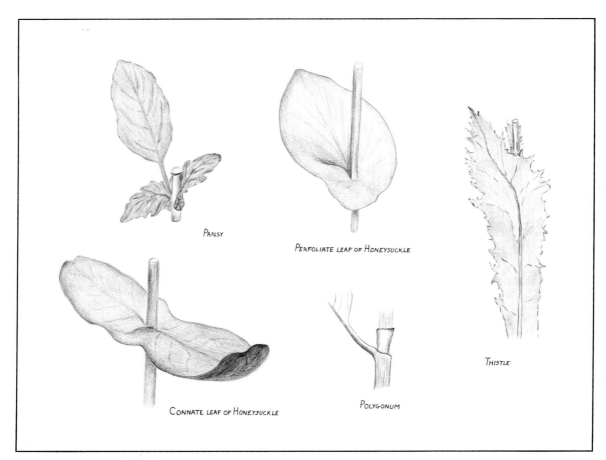

PANSY

PERFOLIATE LEAF OF HONEYSUCKLE

THISTLE

CONNATE LEAF OF HONEYSUCKLE

POLYGONUM

87   Various Forms of Leaf-Base

88   Teasel

A picture of this kind often conceals the fact that the two entities into which the one divides – or which unite into the one – though outwardly alike, may be of quite different character, perhaps even in polar relation to one another. Nature indeed often hides her polarities in seeming symmetries, such as the superficial symmetry of left and right in man, just as she often conceals an underlying unity of type in seeming diversity. So it is with this Cassinian picture of the dividing of one into two if we apply it to the severance of the petiole from the main stem, or maybe actually see it in cross-sections in such a case. For convenience, we will imagine the cross-section so oriented that the petiole is below and the stem above. In effect, the petiole and the main stem will be to one another as positive to negative. The petiole is predominantly physical; the main stem with the "spiritual staff" at its focus represents the ethereal pole. The Cassini-forms in Plate XVI serve also to suggest the nature of this polarity; the lower and narrower region of these forms will be occupied by the petiole – in Nature often semicircular or crescent-shaped, as if to indicate that it is part of a larger whole – whereas the upper and more ethereal part embraces the main stem and the spiritual staff in its midst. Nor does this always remain merely ethereal, it very often becomes a physical phenomenon. The base of the petiole widens into a sheath, more or less fully embracing the parent stem. In the Umbellifers, such as the Hogweed or Cow-Parsnip pictured in Fig. 12, a deep vagina is formed, from which the young shoot springs. In the later stages of development, the flower-bud appears from within this sheath, like a child in the shelter of its mother.

This sheathing or embracing quality, which may be assumed by the leaf-base, reveals a significant aspect of the leaf's potential, as we shall see when we watch the Proteus making the necessary metamorphoses, which will bring the plant to its culmination (paragraph 55). The sheath can be seen in many forms – in many sessile leaves, but also in leaves with long petioles. It appears in the stipules of Pansy (*viola arvensis*), in the connate leaves of Honeysuckle (*Lonicera periclymenum*), for example, or again in the ochrea of *Polygonum* (Fig. 87). A beautiful example is the Teasel (Fig. 88). One feels that this realm of the leaf-base is very important and worthy of a more detailed study. There is a tendency of the leaves higher up the stem to become more sessile and for the leaf-base to become more developed. The petiole, on the other hand, becomes less and less outspread and appears sometimes to dwindle away altogether. This can be seen beautifully in the Figs. 84 and 85, Scabious (*Knautia arvensis*) and Wall Lettuce (*Lactuca muralis*).

Chapter VIII

THE WORLD OF THE FLOWER – FULFILMENT

51 Culmination of Mercury's Staff

When the plant comes to flower, the unlimited progression whereby it rose towards its sun-realm or "star of life" is brought to an end. Functionally, the star-centre has been acting as an infinitude, determining a seemingly endless series. Yet Nature brings this to an end in the flower, allowing the infinitude, or the ethereal realm whose presence it reveals, to act in another and more qualitative way. "Nature", says Goethe in effect, "ceases her growth and closes her account, thus cutting short the possibility of going on in single steps *ad infinitum*, in order that she may more speedily reach her goal through the formation of seeds."[67]

All the time the Mercury staff was in the process of development, to look into its growing-point was to look into a world full of promise. The beautifully arranged, delicate forms, so often revealing the infinitude within of the spiral of life, told of a future state, towards which the whole plant was striving. The flush of colour, so often to be found in the growing point, gave a hint of what was to come; outspoken is the beautiful blue or scarlet of the terminal tufts of coloured bracts or leaves of *Salvia*, which steal all the glory and leave the plant with an insignificant flower in return.

When the tremendous contraction takes place, preparatory to the development of the flower, the leaves, instead of appearing at intervals up the stem, are suddenly ringed around a single point. To form the calyx, the leaves – now metamorphosed into sepals – are gathered and often joined together to provide a covering for the bud and a base for the open flower. The White Campion (*Lychnis Vespertina*) in Fig. 15 is an example; other examples are shown in Plate XVII in the Rose and in Plate XVIII by the Meadow Clary (*Salvia pratensis*), and in Fig. 89 (top left to bottom right) by the Plantain floret (*Plantago major*), Dead Nettle (*Lamium*), Common Mallow (*Malva sylvestris*), and Harebell (*Campanula rotundifolia*).

In most of the higher plants it becomes very evident that here a new phase is reached: from now onward the leaf, that master builder, will perform miracles

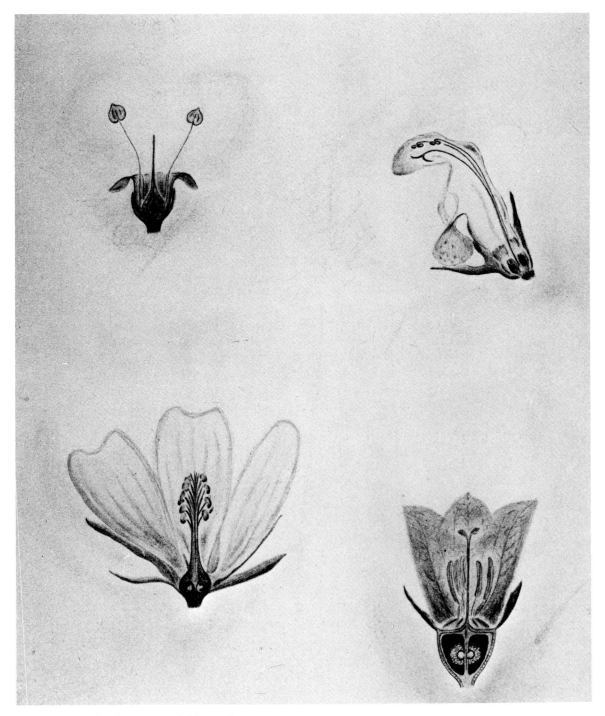

89   *World of the Flower – Cassini Space-formations*

of metamorphosis in order that the aim of the life of the plant may be fulfilled. Among the myriad ways in which Nature reveals her secret in outer form and colour, the Wild Rose (*Rosa canina*) (Plate XV) is perhaps her ideal. The Arnica (*Arnica montana*) in Plate XIV, healer of wounds, with its association of individual florets, has a special virtue.

The final stage in the life-cycle of the plant is now approaching. The essence of it will be that the two polar sources from which the plant sustained its life will become one again. The morphological polarity, called forth with the germinating of the seed – with the first sundering and expansion of the bipolar space described in paragraph 45 – this will once more give way to the single and concentric nature of the new year's seed. Such is the ultimate process, the different phases of which we have now to indicate. The earthly and the sun-like centre will become united.

Spiral progression of leaves and nodes having come to an end or become condensed into a circle in the calyx, the metamorphosed leaves now gather around the star-centre. The calyx, as its name indicates, tends to the forming of a single cup; in the Sympetalae the petals too join their edges as if to emphasize the ether-sphere which they envelop. These metamorphosed leaves no longer unfold and come away so freely; they tarry here, circle within circle. The plant is about to come into a different qualitative relation to the "star" or to the archetypal realm of which this is the "infinitude within". Nature now resolves on a complete metamorphosis in her relation to the ethereal and sun-like realm which has hitherto induced the rhythmic growth.

But this is not accomplished in a single stage. Although the flower is as a rule in outer space so small a portion of the plant, it is significant that of Goethe's three periods of expansion and contraction (paragraph 1) two are still to come. The plant does not at once reach the more tangible result in fruit and seed. In the unfolding of the fair but evanescent flower, it first reveals something more of the nature of that immaterial realm, focussed in the "star of life", which will in some way be embodied in them.

The flower-bud, when it appears at the top of the stem, remains folded quietly and often seems in no hurry to open. Many flower-buds make their own typical gesture or movement in the process of opening; the Daffodil bud is poised almost vertically upright at first and then slowly inclines its opening trumpet towards the Earth; the round bud of the Poppy hangs down at first, and then opens its coloured chalice filled with stamens upward to the blue sky. The impression is that in this closed and contracted state the plant must pause awhile, preparing for what is to come. Clearly the stage is set upon which Nature reveals in outer, visible form, something of that mysterious, inner invisible force, which has indwelt and informed the plant in all its stages of development.

When at last the leaf changes to a coloured petal, it becomes more "ethereal" in form and beauty. Midribs and veins – physical bearers of the green leaves – become amazingly delicate and almost vanish, while the fine surface of the lamina still faithfully envelops the ether-sphere. The petal now becomes an organ of the light in a less material, more evanescent sense than before. While the green leaf drinks in the light to assimilate the carbon, using ethereal energies to renew the nourishing and sustaining substances of Earth, this function ceases in the petals. They now reveal the many-coloured glory of the light itself; they are like Nature's counterpart to the eye that sees them, of which Goethe wrote that it is also sun-like, formed by the very light that it might behold the light.[68]

52 *Lemniscatory Spaces of the Flower*

The humblest flower is a world fraught indeed with wonder. Varied as are the shapes of leaves and the typical structures of different plants in their vegetative stages, this is but little compared with the variety of form and gesture of which Nature is capable in the flower. In a quite different way, colour is now an outstanding characteristic of the stage to which the plant has attained. It is true that colour other than the green of plant chlorophyll has already entered in, often with most brilliant effects. Sometimes the red glory of a geranium leaf or the coloured bracts of *Salvia* can rival that of the flower. Significantly, the young unfolding leaves at the growing point are so often delicately tinged with colour, which they are afterwards to lose. It is as though always in the region of the star-centre there were a latent tendency towards what will afterwards come to full manifestation in the flower. Indeed, colour in the vegetative phase always has something of anticipation about it – or it takes hold like a pirate, robbing the riches to come. It is in the corolla, however, that the world of colour comes into its own; this is its true place, and the plants proclaim their individualities in all the tones of its beauty.

The flower still expresses the bipolar character of the entire plant; it is indeed like a greatly transformed plant growing at the top of the Mercury staff. There are of course many forms of flower, and botanists are often at pains to decide what is sepal and what petal. But the great majority of flowers will show a contrast between the more leaf-like, green calyx (or involucre), and the coloured delicacy of the corolla (or corona) and the organs adjacent within it. The polarity is still there between what is related more to the earthly world below, and the ethereal, light-filled realm above (Fig. 89). While the ovary below is obviously the more material and earthly part, the stigma (which belongs to it) is borne upward by the style into a very different region – the coloured chalice enveloping the star-centre. Very revealing are the foliate stigmas – as in Narcissus or Gladiolus – where little leaf-like organs repeat the typical enfolding gesture.

In this bipolar form of the flower, there is a fourfold series of metamorphosed leaves – the sepals of the calyx, the petals constituting the corolla, the carpels forming the ovary in the pistil, and lastly the stamens. Sepals and petals retain more of the leaf-like form and, as they gather into a whorl, form the essential cup or open chalice of the flower. Ovary and stamens are the seed-forming organs – far more radically metamorphosed.

The four may, however, also be paired in another way: sepals and carpels on the one hand and petals and stamens on the other.

Sepals and carpels, retaining more of the green, leaf-like colour, are as though lifted up by the vegetative part of the plant to receive what will now be imparted. When the ovary ripens into the seed-vessel or the fruit, the calyx frequently takes part in this formation, or else it forms a vessel round the vessel. Often these organs are not quite easy to distinguish from the expanded or hollowed receptacle, the uppermost portion of the stem; thus for some plants the botanical definitions have varied from time to time. Calyx and carpels still belong more to what has been raised by rhythmic stages from below, and they last longer.

The petals and stamens on the other hand are profoundly changed in quality and colour, and it is rare for either to remain; with the passing of the flower they fall away or die. They belong more to the immaterial region of which the star-centre is the living focus.

This qualitative distinction sometimes finds spatial expression. The flower itself will then appear as the final stage in the "Caduceus-lemniscate" progression (paragraph 45) – in a lemniscate-form more or less distinct. The lower loop, still to some extent in the region of green leaf and stem, becomes the vessel to contain the seeds, above it is the more ethereal and transitory chalice with the bright petals and the golden stamens (paragraph 57). Often the stigma reaches up into this latter realm. Whatever be the spatial distribution, something of the polarity we have described is at work and will find its ultimate resolution in the descent of the pollen into the waiting ovules.

Passing from green leaf to open flower the plant has indeed crossed a kind of threshold, and we must try to form some notion of what constitutes this sudden change. Our insight hitherto has been guided by the *forms* of space. But the forms are not all. We have supposed that an ethereal space is here at work, determined by the ideal "star" that hovers in or over the unfolding plant. Ethereal space is the negative counterpart of the space of Euclid. For an analogy, let us return for a moment to this familiar kind of space. We can imagine that this space was given, with all its inner law and structure – with its infinitude and its resulting measures, angles, volumes and so on. Yet this would still be no more than empty space. It is not only the form; it is the content that makes the world. Even if crystal shapes were now to be conjured into this empty space, we should behold varieties of formal beauty yet still no more than forms. In the real world as we know it, the same crystal shape can be occupied by many varieties of substance. An empty room of whatsoever form can be filled with many contents, fair or foul.

Transfer the analogy to the idea of ether-spaces. From the phenomena it is quite clear that the ethereal space of the plant, or of its "star", contains an archetype of form. The shapes and phyllotaxis of the leaves reveal it; the cyclic numbering and pattern of the flower even more so. But this is still not the full content, if there be content in the ethereal realm, analogous to substance in the physical. It is the content surely which is more revealed, which is somehow transmuted into physical appearance – colour, scent and fragrance – when the plant comes to flower. Something was latent hitherto, which the plant now is able to drink in. *Something was held in reserve, which the plant now reveals, even in physical form and substance.*

Our geometrical insight at this moment will only tell us in what direction we should look for these new gifts; it is from other realms of experience that we shall learn to know their diverse virtues. Whence is the glory of the flower? If substance is poured out into a hitherto empty room in the space of Euclid, it comes always from some concentrated source; from thence it gradually fills the room, to the periphery. If it is volatile substance, it reaches the containing walls, or if there are none, tends to the infinite periphery of space. This quality of thought, *in polar opposite*, will now be our guide. At the sun-centre of the ether-space is nothing – only the vast infinitude is there. But in the distant bounds of space is the ethereal content, tending to fill *from without inward* the ether-space belonging to this star. As a material gas or solute tends naturally out to the *periphery* of the physical space which it inhabits, so will ethereal virtues, if such exist, tend inward to the *innermost infinitudes* of their respective spaces (paragraph 30).

Only the difference is, as we have said (paragraph 29), as between one and many. The archetypal entities of the ether-world can first create the spaces which they are then to fill and "inform". In physical space, the plane-at-infinity is one, and it is there once and for all, while in the ethereal realm, the archetypes of life somehow have power to sow countless seeds which they inform. The most essential seeds are these ethereal infinitudes, and Goethe's saying "Everything transient is but a parable"[69] is nowhere more true than here, for every physical seed or spore or germ-cell is but a parable – and a potential receiver – of the seeding power of the great archetypes that work in from the peripheries of space. While the physical material comes, as it must do, from the Earth, the ethereal virtues, which become incorporated in the flower and then ultimately in the fruit are of cosmic origin. Specific to the type of plant, the "star of life" and the flower sphere it forms, is receptive only to its own form and content – drawn inward, planewise, from universal space.

The gesture of the lemniscatory space-formation is a most valuable guide in the understanding of the gesture of flower-forms; we see it revealed in the contrasting shapes of carpel and corolla. It is to be seen most obviously in the epigynous type of flower, like the rose, where it appears in the very form of the flower, with the carpel showing the lower loop, while the upper loop is expanded into the flower. The ovary is like a more earthly form below, bearing the chalice of delicately coloured petals – physical and ethereal spaces once more, but raised to this higher level. The gesture reveals on the one hand, the strong and substantial form of the ovary, built to contain the ovules, the potential seeds; on the other, the ethereal delicacy of the petals, their gradually opening chalice hollowed to contain the stamens. This is a realm receptive to the light, akin to it, and ready to receive the vital gold as the stamens release their pollen, ripened in the sun and air. The polarity is still there between what is related more to the earth-water realm below and what lives in the dry, light-filled warmth-realm above; it shows itself in the green convexity of sepals and carpels on the one hand, and the varied world of colour in the flower-chalice on the other.

If the Wild Rose (*Rosa canina*) is a singularly clear instance of the Cassini-Space gesture in the flower (Plate XVII), to the awakened eye every true flower will reveal in one form or another this polar aspect. There are so very many forms. Compare, for example, the laterally formed Cassini-Space in Plate XI with the Meadow Clary in Plate XVIII. (Here, the curves are due to the fact that the inward-moving ethereal circles are related to an *eccentric* point).[62] The transformation of the Cassini-Space shown in Plate XI, and a glance at the flower-forms in Plate XVIII will surely still evoke the idea of a positive-negative (Cassini) space-formation.

The differentiation of inner and outer Cassini-forms helps in the understanding of different types of flower. Every Cassini-curve or -surface relates a lower region where the physical predominates to an upper, more ethereal region. In the curves outside the lemniscate the two are united by a more continuous transition. Inside it, where the single curve appears in two seemingly separate halves, we have an oval within the lower loop, in the realm of the radially formed spheres, and a distinct oval within the upper, where the star- or sun-centre is prevailing. We are reminded of how the delicate petals in many dicotyledonous flowers appear more or less detached, only alighting on the denser and more earthly organs, to fly away with the wind even before they fade. In the lily type, on the other hand, it is as though the petal-forming process embraced the whole lemniscate from without. The petals become more substantial; the earthly and the sun-like, ethereal poles merge into one another more. This helps in the understanding also of the hypogynous flower. An example of the latter is the Waterlily (*Nymphaea alba*) at the bottom of Plate XVIII, which has been drawn within the type of lemniscatory space-formation shown in Plate XI (bottom left).

It must be remembered (cf. paragraph 45) that the Cassini-curves are merely illustrations, ideal models to help to guide our thinking to perceive and begin to understand *a type of ideal process*, the trace or signature of which is left behind as gesture of form in the living plant. As a visible, material form, the plant is an outcome of such a process. The curves simply express in their way the dynamic interplay of two contrasting universal processes, which pervade the whole of space, flowing through it like waves in a continual rhythm of metamorphosis. Momentarily, in Nature's forms, the gesture of such a process is caught and held for us to see, before it changes and perhaps passes away again in the flow of time. This requires presence of mind, clarity of thought and imaginative perception, and a great love of Nature and of Mother Earth. As the plant rises to an expression of worlds beyond the Earth, yet received within her womb, so will science one day arise from the tomb of materialism, learn *to contemplate the living world of plants and thus to see how the Universe beyond the Earth sends in from distances of space the forces which draw the Living forth out of the womb of the Lifeless* (Rudolf Steiner).[10]

VIII
182

In the fully-formed flower, we have seen the four types of "floral leaf", sepals, petals, stamens and carpels. Added to these organs are those which are directly employed in the formation of the seed, namely, the pollen-grains, which appear within the stamens, and the ovules, which are hidden in the carpels. But together with all these there is a seventh organ, the floral axis itself. This sometimes widens into a flat disc or a shallow saucer or even hollows into quite a deep cavity. It is what is called the receptacle (receptaculum), and it is in fact the top of the stem, which forms in one way or another a base upon which the flower is set. Something which in the vegetative shoot is the almost microscopically small growing-point has now grown bigger and changed in order to bring forth the floral organs. Through this, in contrast to the shoot, there is no longer the power of continued repetitive growth. The shoot has a meristematic layer of cells by which vegetative growth is perpetuated; a leaf can become fully grown, but not a shoot! But in the receptacle, the flower-stem attains its specific form and size and in normal growth goes no further (unless later on it has the task of taking part in the ripening of the fruit, when new growth takes place under new conditions).

Just this innermost region may sometimes be the first to stop growing, and is then overtaken by the periphery, so that there then results the disc- or saucer- or, as with the Rose, the cup-formation. Thus, the receptacle upon which the petals grow, may be more convex – more or less parabolic, like the top of the vegetative stem at the growing-point, though usually larger – or it may be flat or hollowed inward. Usually, it is green, though it often has flower-like functions; for example, nectaries may often be found on it, as well as on the petals of a flower.

Looking at the four circles of actual floral leaves previously described, we find that they form concentric whorls (or even closely-wound spirals, as in the Waterlily). Ranged from the outside to the inside are sepals, petals, stamens and carpels, sepals and petals retaining most of all the leaf's archetypal form, except that interestingly enough there is no longer a vegetative bud or eye, as there always was with the true leaf (see paragraph 48). Just as there is quite obviously a close kinship between the outermost and the innermost of the sequence – sepals and carpels, so also there is between petals and stamens. This is revealed through the fact that petals and stamens are so often attached, and in the incipient transitional possibility between the two which is utilized in the cultivation of double flowers, where petals may be formed at the expense of stamens.

This qualitative difference, besides the difference in colour between the calyx-carpel realm and the petal-stamen realm is the one expressed most clearly by the form of the flower. In the Rose, where the receptacle is hollowed out to contain the ovary below the corolla, the sepals of the calyx and the petals of the corolla grow from the narrow opening at the top of the hollow form. Above this place of constriction the petals open upward to form the chalice containing the golden stamens (Plate XV). Here the gesture of the lemniscate itself is most evident. But it is evident, too, in other forms, such as the Harebell (*Campanula rotundifolia*) at the bottom of Fig. 89.

In flowers with a superior ovary, however, this lemniscatory gesture is no less evident, as in the other three flowers in Fig. 89. Here the petals and sepals grow out from below the ovary, but they still envelop the inner realm of the flower and the star-focus (Plate XVIII). The petals form the surfaces characteristic of that part of the Cassini-forms which is related to the star-centre, while at the same time they contain within their sphere the more physical realm represented by the ovary. There is still the difference between the more moist and the more dry pole – the green part and the coloured; and in both cases the crossing-point of the lemniscate is the place of transition from one realm to the other. Flowers are indeed true revealers of the lemniscatory nature of physical and ethereal space-formations.

54   Leaf- and Flower-colours – Green and Peach-blossom

The relationship between calyx and corolla and together with it between green and the colour which is polar opposite to it, called by Goethe "*Rot*", "*Purpur*" or "*Pfirsichblüt*" – the colour of Peach- or Almond-blossom – is of archetypal significance.

The plants are sense-organs of the Earth, like the springs and wells (see paragraph 46). Just as the eye, coming to rest on a leaf-green colour brings forth as an after-image the Peach-blossom colour, so the green leaves, born in the ethereal realm of the sun-space, evoke the ethereal peach-blossom colour, which belongs to this space, calling it forth into visible, physical form and substance. The peach-blossom colour of the Wild Rose is the colour which belongs to the sun-spaces, just as green is the colour of the leaf-bedecked Earth; it is the colour which arises as an activity of the eye itself. Looking at green the eye often perceives peach-blossom – the complementary colour. Goethe describes this process of the eye in looking at the colours, whereby for each colour the polar opposite or complementary colour is called forth and he calls them physiological colours; the eye creates them.[70]

(The reader will perhaps be surprised to learn that the peach-blossom tint, which is to be seen *outside* the leaves in Plate I and is contained in and above the green cone of leaves in Plate II, does not exist as a pigment in the original pictures from which the plates were made. This rose-colour was entirely due to the complementary colour-effect in the eye of the observer. The same applied to the pink colour *around* the Rose in Plate XV, and to the pink above the cone of young leaves in Plate XX. These physiological colours vary in strength as a result of both external conditions and those of the observer. They depend on the activity of the etheric body in the human eye.)

It is true to say that the colour of the Wild Rose is related to the sun-like ether-realms, just as green is the colour of the fields of Earth; it is the polar reflection of the green leaves, which have in their turn been called forth by the activity of the Sun. We are reminded of Goethe's description of the interplay of the light and the dark poles of the colour-circle.[70] "We can hardly refrain from thinking, when we find the two principles producing green on the one hand and red on the other, in the first case on the earthly in the last on the heavenly generation of the Elohim.''

This peach-blossom colour is very different from the more saturated, autumnal, earthy colours of autumn leaves and the many fruits, as for instance, the Rose hip (Plate XVII). One has the impression that the plant reveals itself to begin with in the four colours which Rudolf Steiner, in his theory of colours[71] calls ''Image colours'': peach-blossom and green, black and white. Peach-blossom is, as it were, the archetypal colour of the blossom – the colour of the sun-realm out of which the flower springs. Characteristically, many white flowers show a delicate blush of peach-blossom, e.g. Wood Anemone (*Anemone nemorosa*) and Daisy (*Bellis perennis*), particularly on the newly opening bud. It is to be seen in the veins of the petals, especially on the outer side. But also, when the flowers shine forth in the full glory of the ''lustre colours'', as Steiner calls them – red, blue and yellow – it is as though peach-blossom were their common source; all the flower-colours are, as it were, variations of peach-blossom over against the green of the leaves. Many flowers show what might really be called an archetypal phenomenon, when the young pink of the bud and the young flowers gradually passes over into a deep blue, as for example in the Lungwort (*Pulmonaria officinalis*), Forget-me-not (*Myosotis arvensis*).

We touch here on a realm of future research. Insight into the way the colours of the plants reveal the polarity and interplay of sun- and earth-spaces and processes will one day lead to a more profound understanding of organic chemistry. Something which is revealed to direct observation, which rejoices the soul and speaks to the spirit in the sensible-ethical (*sinnlich-sittlich*) effect of form and colour will provide the key to the astoundingly quick and versatile changes in the living chemistry of carbon and its related elements. These give rise on the one hand to the light and dark green of the chlorophyll, which is bound to the chloroplasts, but also to those strange antho-cyan pigments, which are dissolved in the cell-fluid, and which as peach-blossom and mauve play back and forth between red-blue and blue-red, and often permeate the whole shoot in *statu nascendi*.

Just the fact that every leaf surrounds a sun-space (paragraphs 49, 50) helps one to understand that the green of the leaf, even into its material aspect, is played upon by peach-blossom. How often are these tones to be seen in the mid-rib and veins of leaves, but also at the leaf-base! Thus, for instance, the King-cup (*Caltha palustris*) has a beautiful, shining rose-purple colour on the bracts surrounding the origins of the leaf-stems, hidden deep in the watery slime. Here the sun-force descends into the earthly realm, just as the leaf-blade, which envelopes the sun-like loop of the lemniscatory space, shows forth the green as ''earthly imagination''. Here, too, there is evidence of the reciprocal interplay of the two poles.

The Compositae Family, the largest family of flowering plants, speaks perhaps most eloquently of all about the unifying quality of the inner forces of the Sun, which are received unceasingly into the womb of Earth, wherever life comes into being. The Sunflower (Fig. 90), the Dandelion, the Common Daisy, which grows and spreads so joyfully, if allowed, on our English lawns; many of the healing herbs, Chamomile, Milfoil, for example, and the Thistle are among them. The German language has the imaginative name "*Korb-blütler*" or "Basket-flower" for them. The tiny flowers are gathered together and closely bunched into a composite head, surrounded by the protective involucre, which is composed of a whorl or whorls of metamorphosed leaves (bracts or sepals) at the base of the inflorescence. Each floret is a complete individual – the petal with its stamen, the carpel with style and stigma. The tiny fruit is often carried on the wind by a parachute of hairs and the stems of these plants often contain milky juice. As we have already remarked, the "spiral of life" is revealed by the form in which these florets associate together in the earthly space into which they are born; the reproductive capacity of these very mobile seeds is to be noticed in the spread, for example, of dandelions and thistles.

In the transition to the inflorescence, we have seen (Figs. 84 and 85) the foliage leaves beginning to draw inward from the periphery and to expand where they are joined to the stem, and now we see them gathered together and often fused, to form a base for the flower. This may happen all of a sudden, or it may happen gradually; in the Sunflower (Fig. 90) we see the stem-leaves change and gently creep, as Goethe puts it, into the calyx. Sometimes, of course there is no calyx, as in the Poppy, or it has slipped down the stem, as in the Anemone.

In the formation of the calyx, we are reminded of the cotyledons, where one or two and sometimes more leaf-like organs formed the prelude to the development of the Mercury staff, or of the formation in some plants of rosettes of leaves, close to the soil. There was a threshold to be crossed then, as the plant emerged from the soil and prepared for what was to come. Now, again, the plant is at a threshold. Nature has gathered the nodes and the leaves together, fused or sometimes quite unaltered, and for the most part, she has left them green, ready to receive the gifts which are to come. Nature's baskets will be filled to overflowing!

55 *Metamorphosis from Leaf to Carpel – The Ovules*

Many are the ways in which the leaf changes its form in order to create a vessel or container for the future seeds. Following Goethe, botanists regard the carpels as metamorphosed leaves in the sense that one or two or more leaves have turned inward, growing together at their edges, where they reach in towards the central axis. Grohmann[52] has a fine illustration showing the fusion of three leaves to form the seed-capsule of a Tulip, with its three carpels. In this inner region – often quite evidently at the union of the leaf margins – there arise the ovules. Two examples are given in Fig. 91 Nasturtium (*Tropaeolum*) and Bluebell (*Scilla nutans*). The third example is a most instructive abnormal development of Columbine (*Aquilegia*); the single carpels, still so very leaf-like, should have come

90   Sunflower (from Hortus Eystettensis)

91   Metamorphosis to Carpel

92   Bryophyllum calycinum

VIII
187

together in the middle. They have, however, remained open and apart, bearing on their margins little leaf-buds instead of ovules. The stigma, which is sterile, is reduced to a little glandular head terminating the midrib of the carpellary leaf.

In the lower part of this Figure the Cassini curves shown in Plate XVI are reproduced, this time as though to represent the sections of three leaves gathered around the central axis at one point, their midribs facing outward, their laminae hollowed about the ethereal realm within.

The change from foliage leaf to carpel is radical. The foliage leaves had always opened out and away from the ethereal realm of their origin; now for the first time the leaf-like organ closes inward and unites substantially with the central axis – the "spiritual staff". The junctions of the carpels are often welded together very strongly, and a distinct rib can here be formed, which may be even more evident than the original midrib of the metamorphosed leaf. The hitherto more hollow region of the "spiritual staff" is at last filled, and it is as though when the leaf-edges reach this inmost region they became fertile, producing the ovules, forerunners of the future seeds. *Bryophyllum calycinum* – a plant which greatly interested Goethe – reveals an innate germinating capacity at the leaf-edges; its foliage leaves periodically produce little budding plantlets all round their margins, capable of sending tiny roots out into the open air. These buds drop off and grow like seeds on the soil below (Fig. 92).

Carpels are formed of the union of the leaf with the "spiritual staff", the ethereal pole of the plant's being, and with this union a new fertility is born. The ovule, hidden in the dark cavity of the ovary, is an enhancement of the bud in the axil of the foliage leaf; it is the first stage in the production of the true seed. This transformation of the leaf is more concerned with material substance; vessels of all kinds are formed, later to become fleshy fruits, hard woody nuts and all the variety of seed-vessels which the plant world fashions. Sometimes a persistent calyx shares the function. In the White Campion (*Lychnis vespertina*), Fig. 15, a strong little urn will be formed to hold the seeds, often standing long into the winter (Fig. 17).

It is interesting in this connection to look again at Figs. 84 and 85, which show the transformation of leaves up the stem in such a way that one sees the more physical aspect of the leaf – the petiole and the base – becoming ever more dominant. This is a very common phenomenon. The leaf-base becomes more substantial and the enfolding, container-forming gesture would seem to foreshadow the task which the leaf, in its metamorphosed form has now undertaken. For in the carpel it has now transformed itself in such a way that the leaf-edges, which as we have seen (paragraph 48) are like images of the far-distant horizontal line at infinity, instead of spreading outward have now united with the line-at-infinity within – the verticon of the plant stem. Here the leaf has indeed used its substance in the process of metamorphosis to create the carpel – the vessel for the seeds. "Verticon" and "Horizon" have united! (cf. paragraph 44).

It happens sometimes that instead of developing normally, the metamorphosis is foreshadowed, as in the Tulip seen by Goethe (Fig. 93). In Tulips,

93   *Goethe's Tulip*

94   *Goethe's Proliferous Rose*

95   *Proliferous Calendula*

VIII
189

an organ which is quite evidently a mixture of leaf and petal may appear somewhere up the stem, looking quite out of place in its abnormality. In Goethe's Tulip, the abnormal organ is joined partly to the flower-stem and partly to the flower itself.

Or the metamorphosis may fail entirely, in which case no seed-forming organs are created and an abortive flower is formed, from which a shoot springs forth, bearing true leaves and even further attempted flowers. From a flower-head new flowers may appear from buds that have developed between the flower-leaves or florets. Goethe gained much insight from such abnormal forms; Fig. 94 is from a sketch of a proliferous Rose in his collection. Fig. 95 shows a Marigold (*Calendula officinalis*); the old flower-head, upon which the seeds had developed, bears a whole crop of young flowers.

In normal development, Nature, as Goethe says, sets a term to vegetative growth, bringing it to an end in the flower. She checks the possibility of continuing to the infinite in so many single steps, in order through the development of seeds to reach her goal more quickly.

These proliferations occur in wet seasons and as a result of over-nourishment. The plant prolongs the vegetative phase in a distorted way. The time should come, however, when the forces of the light gain supremacy over the forces of the earth surging upward. The "Staff of Mercury" must come to an end sometime, if true fulfilment is to be attained. The normal flower bears witness to the ever-increasing sovereignty of the light in the life of the plant.

The very fact that the higher plant grows rhythmically between darkness and light, thus reaching the flower gradually, is closely related to the whole phenomenon of metamorphosis. Having refrained from involving the "star-of-life" in its material substance at a too early stage, it now is called upon to do just this. Now "light" is dipping deep into "darkness" and "darkness" has been raised miraculously to the "light". Perhaps, now that the rhythmic time-process is ended, time is revealed in its eternal aspect, past and future being in some way ever-present or even changing places. The plant-world speaks with a gentle voice, yet tells of mighty secrets. As lightning cleaves the darkness of Earth, so an ethereal blossom may spring from the dark wood, forestalling the green leaves, as with some apple-, pear-, and almond- or peach-blossom.

56 Leaf-Petal-Stamen Metamorphosis – The Pollen

Very different is the transformation of the leaf into the stamen; indeed, it is opposite in character to the formation of carpel and ovary. In the petal the typical leaf-form is often no less easily recognizable than in the sepal, only it grows more delicate and unsubstantial. The far more radical transformation into the stamen and its pollen-bearing anther takes place in a great variety of ways, but the change almost invariably involves greater material reduction; the leaf does not gather substance and turn inward to the vertical axis; as a substantial organ it almost completely disappears. It can no longer spread its plane; only the pollen-sacs are left. It is as though the leaf's plane were seared away and the virtue of the "star of life" were scattered, communicated to the multitude of pollen-grains on the anthers ringed around it.

The metamorphosis from leaf to stamen is more difficult to understand, and from our present point of view we do not pretend to have gained insight into the many strange and characteristic forms of the anthers. Yet the picture which the pollen-bearing stamens place before us, seen in relation to the flower as a whole, and above all the form and function of the pollen-grains themselves, tell us in what direction we have to look.

The ovules are the outcome when the leaves, the more peripheral organs of the plant, draw inward to the spiritual staff, which owes its virtue to the "star". Looking now at the pollen on the other hand, the circle of the pollen-bearing anthers, we gain the opposite impression. It is as if the virtue of the star had gone outward, as if a kind of "Saturn's ring" should be formed about the star Though in the region of the spiritual staff, the ovules are still formed, relatively speaking, in a realm of darkness, enclosed in a more earthly vessel. The stamens on the other hand bring forth their pollen in the flower's uppermost cup of light. This is the final stage; we are in the region of the star itself – the anthers gathered round it like a corona.

Just as it is the abnormal growths, which reveal the fact of the hidden spiral process in the normal plant, and also of the metamorphosis of leaf into sepal, and carpel, so, too, in many double flowers – in which there is also an element of abnormality – the secret of the metamorphosis of the leaf into petal and stamen may be laid bare. The Garden Rose, for example, often shows as one follows the petals into the centre of the flower, many transitional forms, stages of deformation, between a true petal and a true stamen (Fig. 96).

In the variation of leaf-forms up the stem (Figs. 84, 85), we saw the leaf-base growing materially, foreshadowing the metamorphosis into the carpel, while at the same time the leaf-blade diminished in size. In the development of the plant up the Mercury staff towards the flower, petiole and leaf-base were gaining in substance, as though in anticipation of the forming of strong earthly vessels to house the future seeds, while the lamina was being reduced more and more to a strip-like form or even to non-existence. Here in the stamen we have the impression even more that the physical leaf has all but disappeared from space. The delicate filament is like a memory of the leaf-stem, and in the shrivelled anthers only the pollen-grains are left.

In the metamorphosis into the stamens the outward material expansion of the leaf is utterly reduced. Already the petals have assumed delicate and unsubstantial forms. Now in the presence of the star-centre, the leaf as a physical form withdraws, it sacrifices its substance as if to the unseen flame of light. It is as though the power of this star had overwhelmed and seared the leaf; it can no longer spread its plane. In the partly formed or partly reverted anthers, such as we find in the double flowers, we can see by the contorted forms how great is this searing and contracting power. Yet the apparent contraction is but the physical aspect of the process. The golden pollen-grain in reality, with its seeding, fertilizing power, is pointlike not only in the earthly sense but above all in the ethereal. It is the potential bearer of the star of life itself – of the all-important infinitude to whose presence the whole life and development of the plant until this moment has been due. Like the ethereal infinitude itself, it is a point of many planes. Thus if the stamen cannot unfold its plane like a normal leaf, it is because it has received, like a touch of fire, the infinitude towards which all the other planes were oriented.

57   Fruit and Seed Formation – A New Beginning

Nature has thus fashioned a messenger, an earthly carrier for the ethereal forces of the universe; and as so often happens Nature multiplies, what is ideally but one, to gain her end. A myriad pollen-grains receive the virtue of the star, and every one of them can fertilize an ovule. Insects will carry them from flower to flower, or they are borne upon the wind. And it is symptomatic that in the distant stratosphere these bearers of a sun-like virtue have been found. Earth offered up her substance to the ether-world to produce the leaf – a planar organ, marvellously fashioned from a point-wise world. Ethereal and sun-like space now gives its virtue to the pollen-grain – a point-like organ endowed with forces of a planar world.

It is the final act in the great drama when at long last the star of life – borne by the pollen-grain – descends to the waiting ovules. Hitherto an immaterial focus dominating the unfolding spheres of leaf and branch or hovering in the open flower, the star is now received into the earthly substance to rest within the living matter of the plant. Earthly substance in living form has been raised by the plant, lifted from the ground in rhythmic steps from node to node, until the flower is reached. There is the vessel of the ovary, with its green colour reminiscent of the more earthly realm from which it has been raised; above it or around, the more ethereal chalice of many-coloured petals and the ring of golden stamens.

The ovary, in the realm characterized by the lower parts of the Cassini surfaces, contains the ovules, enclosed in a green cavity into which the light of the Sun only penetrates dimly, if at all. The world of the ovary is dark and moist, reminiscent of the world below the soil in which the seed once was. Above it the petals open, forming the coloured hollow with the stamens within it, their anthers bearing the pollen. In the midst of these the stigma rises from below, where in the still green darkness – as though in the tomb of Earth, yet high above it – the ovules await the messenger of the light. The pollen is at the very summit of the plant's achievement as it strives towards the light. It is usual in the higher plant for the pollen to ripen in the air and sunshine. The pollen-grain represents the "light" pole of the Cassini-space; it is like a materialization of the ethereal focus of the lemniscate. The plant has raised this tiny particle of its living matter into the realm of the corolla; it is a particle, but no ordinary particle, for it is destined to be the bearer of the most vital forces. We may say: it is no ordinary point – it is more like a point of many planes (cf. paragraphs 17, 34).

96   *Leaf-Petal-Stamen Metamorphosis*

97a   *Pollenization (from Wisniewski)*

97b   *Pollenization (from Wisniewski)*

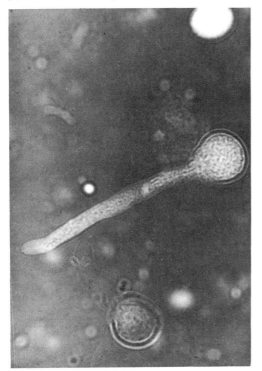

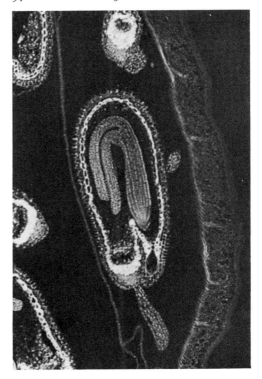

Then comes the moment when in one way or another the pollen unites its substance with the ovules. This happens in what would seem to be a remarkably physical way. A grain of pollen alights on the stigma and sends its pollen-tube down the style to penetrate the ovule. Figs. 97 (a and b) show a pollen-grain of Shepherd's Purse (*Capsella*) and the receptive realm of the ovule into which it penetrates. Many animal-like qualities are associated with this process, both in the methods of fertilization by the insects and in the forms which Nature produces to attain this end. It is, however, very important to realize that this is no animal fertilization; it is the fertilization of Mother Earth by the cosmic father-process.[72]

The picture of the Cassini-space still retains its significance as an idea expressing this great process in Nature. What does it mean if the "light" focus and the "dark" focus draw together and become one? – for this is indeed what has happened when pollen and ovule unite and the seed is formed.

In the last resort, they become the *united* centre of the two families of concentric spheres (Plate X), the "ethereal" and the "physical", by whose interplay, before their centres united, the bipolar space was for ever being formed. In the moment of union of these two centres, the rhythmical lemniscatory forms will have departed, for the polarity will no longer be there to bring them into being. A point will remain, centre of both the inward- and outward-breathing spheres. A seed is like such a point, seemingly so dead, yet with vital forces indwelling it.

The final union of the two poles, resolving the bipolar form of the growing plant into the seed, is indicated in Plate XIX. In a Cassini-space, the relatively distant surfaces approximate to the form of spheres, though the exact form is not reached until the plane-at-infinity with its imaginary circle, which is no longer a true but a degenerate sphere. The realization, in a bipolar living form, of these more distant surfaces will therefore tend to emphasize the organic unity of the two poles, just as the double-oval forms – inside the lemniscate – lay stress on their polarity and difference.

The ultimate fusion of the two poles corresponds to the following mathematical idea. The passage outward from curve to curve in a single Cassini family (Plates X or XI, top left, for instance) signifies the variation of a parameter – *a* in equation (iii), Note 62. Now one can take as a second variable parameter the distance apart of the two foci, reducing this while the former parameter increases, so that the one becomes zero when the other grows to infinity. There are of course many functional relations whereby this can be achieved. A curve-family is then obtained, tending from two distinct foci, with the double ovals and the lemniscate between them, to a finite sphere, which latter represents the moment when the two have coalesced in one. Such, we conceive, is the ideal process in the transition from the vegetative and flowering stage to what is finally achieved in "fertilization" – in Goethe's third "expansion and contraction", culminating in the spheroidal form of the fruit and in the single living centre of the seed. The process is suggested in the progression of Cassini-spaces drawn on an ever smaller scale in Plate XIX.

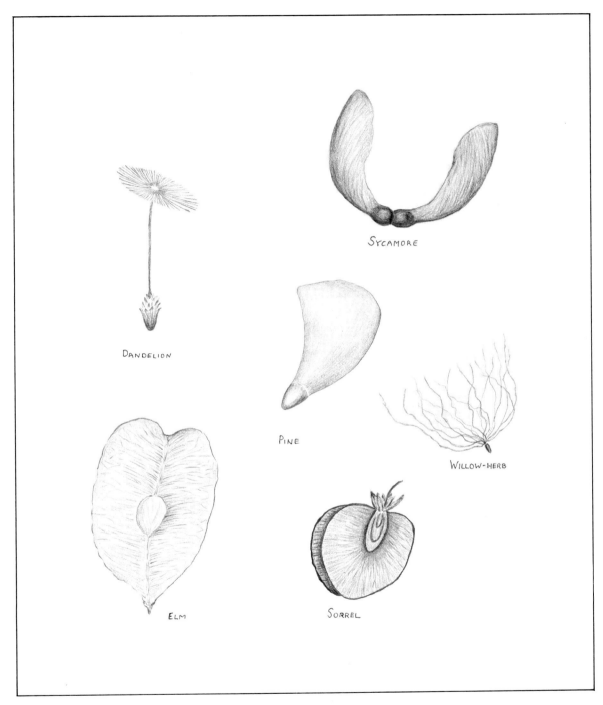

DANDELION

SYCAMORE

PINE

WILLOW-HERB

ELM

SORREL

98   *Winged and Plumed Seeds*

How different is the expanded roundness of the fruit – apples, pears, plums and all the rest, with their juicy sweetness – from the hard convexity of the root, reservoir of salty mineral substances. It is this summer's sun which gleams from the shiny surfaces of the rosy-cheeked apples, the golden yellow marrows, and the many fruits piled on the table of the Harvest Festival. Within them the seeds are hidden, which have received into their very substance this summer's sun, out of the hidden power of which the new plants will be conjured forth next year.

The resting seeds are organisms in which the foci of concave and convex growth – the more sun-like and the more earthly principle of form – are united. In the multitudes of shapes which the plants bring forth as seeds, it is sometimes the pointwise, substantial aspect which speaks predominantly, as for example in the hard and woody seed and seed-capsule of the Brazil Nut, pictured in Fig. 18. Very often the seed, endowed with all manner of planes and wings and parachutes, is calling to mind the planewise world of the light to which it belongs; so often too the spiral is beautifully revealed (Fig. 98 and the Thistle seed in Plate XIX).

The fruits and seeds are garnered now, in the cool and dry places, in their hundredweights, in sacks and heaps. They will provide food for man and beast through the dark winter; comparatively few, the finest specimens, will be needed for the sowings of the coming year; received once more into the moisture and warmth of the Earth and into the rhythms of the Universe, they, with Persephone free, will arise again; the promise of the rainbow will be fulfilled once more, and the future of Nature herself and of mankind will be assured.

58   Afterword

Throughout this book the emphasis has been on the morphology of the growing plant. Chemical and physiological aspects have only been touched on incidentally, as in considering the relation of the plant to the Sun. The question will obviously arise whether this way of approach will also shed light on plant chemistry, and therefore on the great problem of our time – human and animal nutrition and the conservation of the soil. Although this is beyond our present scope, an indication may at least be given of the possibilities which, we believe, will open out also in this direction. Here too will have to lie the proof – not only in contemplation of Nature's forms, but in experiment and practical application.

From the whole concept here developed – the interplay of centric and peripheral formations – it may well be expected that living forms, seen as a whole and penetrated with this kind of understanding, will also provide essential clues to the nature of the substances which they produce and to the chemical formative processes which are today interpreted as an outcome of the minute molecular constitution. There is already, in modern biochemistry, a tendency to overcome the old additive habit of thought which would regard the organism as a mere summation of the molecular events going on in each single cell or volume-element within a cell. Biologists are amazed by the subtle and pervasive quality of the chemical influences – hormones and auxins for example – with which the living organism regulates its life and growth; also by its selective power with regard to osmotic and ionic processes which in the non-living world appear to follow a far simpler course. Recent developments in physics may be relevant to this; some speak of "biophysics" as the future science. But the resources of an "organismic" outlook are reinforced in quite another way – in more direct relation to the whole – by the exact morphological principles derived from projective geometry, as for example by the idea of ethereal or planar spaces.

In one direction we believe that the plant as here seen "between Sun and Earth" will open to the scientific mind a new perception of the realm of substance and of the true life-giving forces of nutrition. Two interlacing principles have been discerned in the bipolar form of the living plant. Polar and complementary to one another, these two principles of growth, which in their morphological aspect have here been recognized in a new and fundamental light, are obviously related to the dual stream in the physiology and physical chemistry of plant life. On the one hand there is the upward transpiration-stream. Carried in the xylem vessels, vast quantities of water are drawn outward from the earth, evaporating through the leaves into the air – plant-life contributing in this way to an essential meteorological process. The other stream is in- and downward. Carbon dioxide from the air is transformed into carbohydrates, building the body of the plant in cellulose and lignin, accumulating in the form of starch, and carried in solution, most probably through the phloem, into the stem and root and other organs where the plant will store or spend or otherwise transmute them.

Present-day chemistry and physics follow these inward and outward streams only up to a certain threshold – a threshold guarded, until quite recently, by the conviction that the conservation of mass was an unalterable law. The concept of ethereal or planar space as well as pointwise, and of the interplay of spaces positive and negative, suggests that in the living world the interrelation of matter and vital energy goes beyond this threshold, namely that in the outspread leaves new, virgin substance is precipitated, as it were, into the pointwise realm of matter. The "star of life" – or the ethereal space whereof it is the focus – is an organ of the plant, a receptive organ at this more sunlike pole of the bipolar form. Plants live in association with one another, taking what they require not only from their immediate surroundings and from the soil beneath them, but breathing-in from cosmic realms. For this activity the star-centres are ethereal organs. A highly refined, potentized activity goes on in relation to these organs, and what is thus received has qualitative virtue; it is no mere quantitative increase. At the same time it is suggested that in the up- and outward stream – within the plant, and it may well be, for the water vapour, in upper reaches of the atmosphere – some earthly matter is etherealized, escaping altogether from the pointwise realm. If this be so, the lemniscatory interplay of inpouring and outpouring activities contains the key not only to the morphology of plant-form but to a process of transmutation whereby a constant regeneration of the Earth is going on – a raying-forth into the cosmos and a receiving from thence anew.

There is a clear mathematical foundation also for this idea of annihilation and rebirth, or of a passing to and fro between pointwise and planar realms. The contrast is acutely felt – Goethe, Ruskin, Eddington and many others have borne witness to it – between the quantitative character of exact science and the rich qualitative impressions man receives from Nature, which in a man of experience, a farmer for example of the old school, turn into practical wisdom. The contrast becomes less sharp when the mathematical approach, without losing its exactitude, enters the more imaginative realm to which the new geometry has opened out the way. There is in fact another threshold which the pure mathematician has long since learned to cross, but which has rarely if ever been crossed in the prevailing mathematical theories of outer Nature – the atomistic theories above all. It is the threshold of the infinite – or of the zero, which is the same in the reciprocal aspect. In pure geometry it is well known that the great metamorphoses are brought about by the crossing of such thresholds. A finite form becomes infinite or zero – maybe degenerating into point or plane – and is reborn in another shape. To take another instance: two or more distinct entities fuse into one, and at this moment the form and function they determine is radically changed. There is more difference mathematically between the final fusion and the minutest distance apart of two points than between the latter and the greatest finite distance. Nay more, the coalescence in the centre is exactly like the coalescence at infinity, when the distant point to the left becomes identical with the distant to the right. The real, qualitative difference is not between great distances and small but between any distance whatsoever and this final fusion.

To think in terms of point and plane – or of infinitude within and not only infinitude without – opens the mind to the fact that in real Nature these qualitative processes,[73] known to the pure mathematician for at least three generations past, are taking place before our eyes. And it suggests that this applies also in the realm of substance. In scientific thought we have to get beyond the stage where finite measurable entities must always hold their own or else give way only when other finite measurable forms immediately take their place. We have to recognize the thresholds of transition where in the centre or in the vast periphery – in point or plane – cosmos dissolves into chaos and as a well-knit form is sacrificed completely, so that new cosmos may be born out of receptive chaos.

In their instinctive way, people in times past were well aware that the plant draws not only from the Earth and from its physical surroundings but in its ordered rhythms of life receives from the universe of Sun and stars; they husbanded their land accordingly. Along the lines here suggested this too may become scientific knowledge, giving much-needed guidance to those who feel the need to treat both plant and soil in the way a living entity deserves. Experience has shown that disappointing and even destructive results may be obtained when the powerful methods of modern chemistry are applied directly to the living world. Greater and greater care is being exercised in this respect. Arising out of such experience, the need is felt for a more integral approach. The molecular pictures of chemistry are too remote from what is seen and known in the immediate contact with Nature.

The tentative idea here put forward concerning the *substantial* function of the life of plants will, if confirmed, shed a new light on questions of nutrition, for all earthly creatures. Not only the archetype of form is of cosmic nature, but the very substance by which the creature lives is renewed and regenerated, inward from the periphery, from the celestial universe.

While these initial suggestions may need to be greatly modified, this much is certain. When the polarities of the spatial universe – expressed in the geometrical Principle of Polarity (Duality) – have duly penetrated into the thoughts and imaginations of science, a cosmic outlook will arise, which will also lead to a new sense of responsibility towards the life of the Earth-planet.

*   *   *   *   *   *   *

These words, written in 1952, are valid now as they were then. There have been far-reaching changes in the world in recent years, as a result of the enormous technical advances made by physics and also by bio-chemistry and bio-physics. By and large, the direction taken has strengthened the materialistic view. On the other hand, aspects of research into areas to which orthodox science turned a blind eye have gained more attention. Serious work is being done in many areas bordering on the limits set by science, the results of which cannot all be brushed aside so easily. Books of a journalistic nature have appeared in increasing numbers, in which reference has also been made to Rudolf Steiner and the work of his school.

There is much confusion and distortion of the truth, but it cannot be denied that there is a prevailing urge to seek beyond the limits set by a purely material-istic scientific method. Nevertheless, until there can be clear, thoughtful discrimination between truth and error, conceptions arise which are even more materialistic.

The question is, however, being asked in biology, for example: Are not the origins of what seems to be going on within the minute confines of the cell, to be sought in some field of forces outside the cell – even outside the organism itself? Such questions are justified, and it is natural that in attempting to find answers to them, the prevailing concepts of physics, such as geo-magnetism, electro-magnetic fields, and the like, come to hand. It will, however, not be adequate, simply to equate the etheric formative forces with the electro-magnetic. Science has uncovered the sub-natural forces and it will one day recognize the truly super-natural, but the two are not identical. The question might well be better formulated. May not forces belonging to a higher realm of being reveal themselves in nature, just as the spirit of a man reveals itself on Earth in the guise of the physical body? Then, however, the question will be: What is the true nature of the form of man?

In this morphological study, we have begun to build a bridge in thought between natural science and spiritual science on the basis of the bridge already built by eminent thinkers in the realm of modern geometry and mathematics. The laws of Earth-space and finite form no longer dominate as concepts; we have acquired concepts and ideas more suited to the laws of the light and of life. Thinking has become a truer interpreter of living phenomena; not only have relationships between things become more visible, but also the possible relationship between the sense-perceptible world and a world not so immediately visible to the senses. The key lies in the spiritual activity of objective thinking and not in mysticism. To awaken a perception of the negative, planewise aspect of space and of living forms is an activity of spiritual imagination based on thinking; it is art as well as science.

The new science of ecology today is gaining ground in the study of the way organisms and different realms of nature relate to and influence one another. It is a good step in the right direction. A next step will be in the true understanding that *Small is Beautiful* – to use the terminology of a modern "best seller".[74] The reality lies not only in the expansive and extensive aspects of life, but also in the less visible realms, which are all-important. If modern science and the culture which rests upon it will seek and begin to find a counterbalance to the all-prevailing, one-sided materialistic idea of expansionism, it will provide the ground for a more truly sociable society. There are signs that we are at the beginning of this road in the world at large.

In comparison to the vast strongholds of science today, the organic movements in medicine and agriculture are small and struggling to maintain their foothold, but there is a growing public awareness. The methods of homoeopathy and organic farming, and still more the scientific methods and research inspired by Rudolf Steiner, suffer the disadvantage that the results of their experiments require *qualitative* rather than quantitative evaluation, which is extremely difficult to achieve – some would say impossible. How can a particular quality be expressed in terms of quantitative measurements?

Rudolf Steiner suggested two possible methods to demonstrate the working of the etheric forces; crystallography and chromatography (Capillary Dynamolysis).[12] Years later, Schwenk devised a third – the "Drop-Picture" method.[51] All three test methods are arranged so as to allow forms to arise on a surface or plane, for, as Steiner had explained and as we have tried to show in this book, the etheric or ethereal forces have a *planar* quality, and it is therefore likely that they will reveal themselves to some extent in planes and surfaces and not primarily in the pointwise and extensive aspect of spatial forms. In the three Figs. 99, 100 and 101 we include examples of experiments in which the substances involved have created a picture, which quite evidently reveals both radial and peripheral qualities in its forms. The task, then, is to learn to read what these forms are expressing.

As a fitting conclusion to this book, prefaced as it is by Ehrenfried Pfeiffer, we also include an unwitting demonstration, which is of a similar nature. (It comes to hand as we go to print and has not yet been further investigated.) Pfeiffer developed a bio-dynamic preparation, which he called "The Bio-dynamic Compost Starter", designed to start and accelerate bacterial processes in refuse, compost and manure on a large scale. Recently, in the laboratory, which continues this work,[75] it has been noticed while packaging, that fine particles had separated out from the bulk of the material and become deposited on the inner

99   The Formative Forces Revealed in Crystallization. Bean-extract with Copper Chloride (Selawry)

101   The Sensitive Surfaces in Water (Schwenk)

102   Forms on the Surface of the Container of a Bio-Dynamic Preparation (Pfeiffer Foundation)

100   Capillary Dynamolysis. Mistletoe Juice with Gold Chloride (Fyfe)

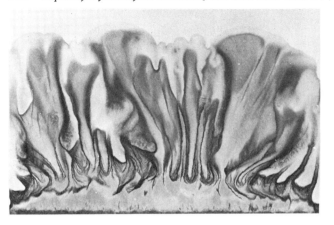

VIII
201

surface of the plastic containers, in forms which certainly exhibit some sort of organizing or formative principle (Fig. 102). The material in this container was no ordinary soil; it had undergone processes designed to intensify the vitalizing and fertilizing qualities of the agricultural preparation. The question at least is justified: Is not a living, plant-like quality actually expressed in the delicate forms, which have in some as yet unaccountable way made themselves apparent on the surface of the container?

A final word remains, concerning work that is in direct continuation of George Adams' researches. In 1959, not long before his death in 1963, he founded with Dr. Alexander Leroi and Dipl. Ing. Theodor Schwenk the institute in the Black Forest, Germany – *Institut für Strömungswissenschaften im Verein für Bewegungsforschung* – which is concerned with research into the qualities of water and into methods for the purification and even rejuvenation of water.[51] The attempt was begun to find ways of moving water, so that its innate spiralling and surface-creating qualities might be enhanced.

Apart from the continued work of the Institute, attention should be drawn to the work of two men who were at that time students and helpers of George Adams – Lawrence Edwards and John Wilkes.

Lawrence Edwards,[64] longstanding student of George Adams and teacher of projective geometry, has continued to open up the field of projective, planar surfaces (Path-Curve Surfaces), which Adams was developing in connection with his researches with Schwenk into water. Edwards has used this as yet untried field of morphological research in regard to plant and other organic forms. Most significantly, he has succeeded in demonstrating mathematically in detail the positive-negative space relationship between the ovary-bearing part of a flower with the bud-form of the corolla of that particular flower, seen as a receptive and transforming Sun-space. The work shows in more than thirty cases of flowers of wild plants that the flower-bud grows in what may be called an active "field of form", containing an "infinitude within", which functions in each case as a transforming sun-centre, relating the cosmic, ethereal or planar processes with the detailed formation of the living material substances in the ovary of the plant in question.

John Wilkes,[76] whose task it became as a sculptor to use his unusual skill in building laboratory models of the Path-Curve surfaces designed by George Adams, continued empirical investigations after Adams' death. He was seeking to relate the moving element water to surfaces within vessels of specially pro-portioned systalic form, and discovered what he subsequently called the "Flow-form Method". This provides the means of inducing rhythmically swinging movements in streaming water, which in turn may be led in a lemniscatory flow-path over surfaces under investigation. This work is attracting wide attention.

Today it is man himself who is responsible for the continuation of his world. Science, which, to remain true science, must be based on the free and objective spiritual activity of man – the creative power of thinking – will open up her fields of research to include the eternal truths, which always have and in the future always will sustain life.

NOTES AND REFERENCES

CHAPTER I

1 §§ 1, 14. Goethe. The five volumes of the scientific works of Goethe were edited by Rudolf Steiner in 1883 for Kürschner's Nationalliteratur (Stuttgart) (New edition, Dornach 1975). Rudolf Steiner's introductions and notes were collected and published in *Goethes Naturwissenschaftliche Schriften* (Dornach 1977), and a selection in English: *Goethe the Scientist* (New York 1950). An English translation of Goethe's *Metamorphosis of Plants* was published by the Chronica Botanica Co., Waltham, Mass. (London 1946), with an introduction by Dr. Agnes Arber. An earlier translation by Olive Whicher has since been published by Bio-Dynamic Literature (Rhode Island 1978). A complete translation of all the work on botany, entitled *Goethe's Botanical Writings* by Bertha Mueller was published by University of Hawaii Press in 1952. Bio-Dynamic Literature has published a small volume, compiled by Herbert H. Koepf and Linda S. Jolly, entitled *Readings in Goethean Science*, which contains several of Goethe's short writings, including the poetic essay "Natura", translated by George Adams, and also Rudolf Steiner's essays on *Inorganic Nature* and *Organic Nature*, taken from his book *A Theory of Knowledge Implicit in Goethe's World Conception* (Spring Valley N.Y. 1978).

Whereas Linné and his followers, in classifying plants, places emphasis on the outer distinguishing marks, such as size, number and position of single organs, in the same manner in which one orders a number of inorganic forms, Goethe's insight into the nature of plant life goes deeper. In his *Metamorphosis of Plants* he recognizes a superior, all-embracing principle, which he calls the "Ur-pflanze" or "Archetypal Plant". He sees it coming to expression in all the different outer forms and manifestations of the plant kingdom; it is that "Something", which makes a particular being of nature *a plant*, and which *unites* all the factors which Linné seeks to separate out.

Rudolf Steiner in his introduction to the *Metamorphosis of Plants* writes: "A living organism is a self-determining whole, which produces its states of existence out of its own nature. Both in the juxtaposition of the parts and in the time-sequence of its states of existence, a living being undergoes a constant interchange of relationships, a reciprocity, which does not manifest as though determined by the sense-perceptible qualities of the parts, nor through the mechanical-causal determination of the later by the earlier, but is ruled by a higher principle, which stands above the parts and the states of existence. It is in the nature of the whole organism that a particular stage of its existence should be the first and another the last; also the sequence of intervening stages is determined within the Idea of the whole. An earlier stage is dependent on a previous one, and vice versa; in short, in a living organism, the *evolution* of one form out of another, the transition of stages of existence into one another, is no finished and rounded-off individual existence but a constant *becoming*. In the plant, this determination of each single organ by the whole comes to manifestation to the extent that all organs are built upon the same fundamental model. . . .

"As the forces which organize the nature of the plant come into actual existence, they take on a series of structured forms. What was now needed was the living concept which united these forces backwards and forwards.

"When we consider Goethe's theory of metamorphosis, as it appears from the year 1790, we find that for Goethe this concept was that of alternate expansion and contraction. In the seed, the plant formation is most intensely contracted (concentrated). With the leaves, there follows the first unfolding, expansion, of the formative forces. What in the seed is pressed together into a point, spreads apart spatially in the leaves. In the calyx, the forces draw together again around an axial point; the corolla comes about through the next expansion;

stamens and pistil arise through the next concentration; the fruit through the last (third) expansion; whereupon the whole force of the plant's life (this Principle of the entelechy) conceals itself again in the most highly contracted state, the seed. Although we have been able to trace fairly well all the details of the idea of metamorphosis up to its final utilization in the paper which appeared in 1790, it will not be so easy with the concept of expansion and contraction."

2 § 6. "Hypocotyl" is the name given to the region where the root passes over into the stem or shoot, which is clearly recognizable in the anatomic structure. According to the meaning of the word "under the cotyledons", it forms the boundary between the radicle and the plumule, already in the germinating seed.

3 § 6. An apparent exception would at most be in the formation of thorns (e.g. Whitethorn; Sloe). But even here it is only possible for the plant to metamorphose some of its growing-points to thorns, leaving leading growing-points to develop normally, in accordance with their function as centres of vital growth.

4 § 6. The horizontal section of a primary root shows the star-shaped arrangement of the wood-fibres (the so-called xylem), which pictures a radial formation, raying out from the centre; between the rays are the bast-fibres (phloem), embedded in the parenchyma (fundamental ground-tissue). In the stem or growing shoot, on the other hand, the vascular bundles, composed mostly on the outer side of phloem and on the inner of xylem, are arranged peripherally and in the main side by side in an outer ring. Here the xylem and the phloem are located tangentially to one another, parallel to the surface of the stem. At the hypocotyl there occurs the complicated rearrangement of the radial formation into the peripheral, which results in an absolute inversion or polar exchange of the inner with the outer. (To be recommended for study in this specialized realm: *Plant Anatomy*, by Katherine Esau (New York, London, Sydney 1965), *Botany of the Living Plant*, by F. O. Bower (London 1947). *Plant Form and Function*, Fritsch and Salisbury (London 1967).

5 § 11. Ernst Haeckel, *Generelle Morphologie der Organismen* (Berlin 1866). Haeckel shows to begin with how in the inorganic sciences, for example, in crystallography, the recognition in pure thought of the possible crystal systems and classes of symmetry contribute substantially to the understanding of the natural phenomena, and he states that for the understanding of morphology in the living kingdoms of nature it is equally indispensable to reach a general view concerning the possibilities of form contained in the whole structure of space. If the fundamental thesis of the present work proves to be true, namely, that for living forms not only the positive (Euclidean) space, but also the varied interplay of positive and negative space-dimensions be taken into account, then this theoretically justified ideal will become greatly deepened and widened.

CHAPTER II

6 § 13. Contrary to some expressed opinions, Goethe, in his natural scientific striving, was not averse to the true mathematical spirit. Werner Heisenberg (in *Wandlungen in den Grundlagen der Naturwissenschaft* (Leipzig 1935) English: *Problems of Nuclear Science* (London 1952), writes: "What Goethe renounced is really not mathematics itself, but only the way mathematics is used."

7 § 13. See among others K. v. Neergaard: *Die Aufgabe des 20. Jahrhunderts* (Zürich 1940). Neergaard cites P. Jordan: "We should like to see in the biological laws . . . as opposed to the inorganic, something more general and all-embracing"; accordingly, "the laws and phenomena of the inorganic world would be a simplified special case of the organic." See also: B. C. Goodwin: *A Cognitive View of Biological Processes* (Journal of Biological Structure, 1978, 1, 117–125).

8 § 13. A. Gurwitsch, *Abhandlungen in Roux' Archiv für Entwicklungsmechanik*, Vols. 51, 52, 101, and particularly 112 (1922–27).

9 § 14. See the following fundamental works of Rudolf Steiner: *A Theory of Knowledge Implicit in Goethe's World Conception* (New York and London 1940); *Goethe the Scientist* (New York 1950); *Occult Science – An Outline* (London 1979); *Knowledge of the Higher Worlds. How is it Achieved?* (London 1976); *Goethe's World Conception* (London 1928).

10 §§ 14, 52. Rudolf Steiner: *Anthroposophical Leading Thoughts*, p. 15 (London 1973).

11 § 14. In his essay on Goethe's morphological writings, Rudolf Steiner describes how in 1784 the poet read Spinoza with Frau von Stein, and he writes: "The influence of this philosopher on Goethe now became very great. Goethe himself always recognized this clearly. In the year 1816 he wrote to Zelter: 'Except for Shakespeare and Spinoza, I should not know that anyone who has passed away had exercised such an influence upon me as Linné.'" (*Goethe the Scientist.*)

12 § 14. A. Fyfe: *Moon and Plant Growth* (Arlesheim 1974), *The Mistletoe in the Cycle of the Seasons* (Reprint, British Homoeopathic Journal, London 1969), *The Signature of the Planet Mercury in Plants* (Reprint, British Homoeopathic Journal, London 1974), *Die Signatur des Mondes im Pflanzenreich* (Stuttgart 1967), *Die Signatur Merkurs im Pflanzenreich* (Stuttgart 1973), *Die Signatur der Venus im Pflanzenreich* (Stuttgart 1978). Wilhelm Pelikan and Georg Unger: *The Activity of Potentised Substances* (British Homoeopathic Journal, London). Th. Schwenk: *Grundlagen der Potenzforschung* (Arlesheim 1954); see also Note 51. M. Enquist: *Physische und lebensbildende Kräfte in der Pflanze* (Frankfurt 1975); *Die Steigbildmethode* (Frankfurt 1977). A. Selawry: *Die Kupferchloridkristallisation* (Stuttgart 1957). E. Pfeiffer: *Sensitive Crystallization Processes* (New York 1975). F. Bessenich: *Zur Methode der Empfindlichen Kristallisation* (Dornach 1960). G. Wachsmuth: *Die Aetherischen Bildekräfte* (Dornach 1927); English: *Etheric Formative Forces* (London and New York 1932); *Die Aetherische Welt* (Dornach 1927). J. Bockemühl *et al.*: *Erscheinungsformen des Aetherischen* (Stuttgart 1977). E. M. Kranich: *Die Formensprache der Pflanze* (Stuttgart 1976). H. Koepf *et al.*: *Bio-Dynamic Agriculture: An Introduction* (New York 1976). *German: Biologische Landwirtschaft* (Stuttgart 1974). The early pioneer work was done in close association with Rudolf Steiner by E. Pfeiffer and E. and L. Kolisko (whose writings are being republished by Kolisko Archive Publications, Bournemouth), and by Dr. Rudolf Hauschka: *Substanzlehre* (Frankfurt 1951), *Ernährungslehre* (Frankfurt 1970).

Continuous research is being carried out in the laboratories of the two pharmaceutical firms, Weleda (Arlesheim, Switzerland and Schwäbisch-Gmünd, W. Germany), and Wala Heilmittel Dr. Rudolf Hauschka (Eckwälden/Bad Boll, W. Germany), also at the Ita Wegman Clinic and the Lucas Clinic (Arlesheim, Switzerland). See also the collection: *Potenzierte Heilmittel* (Stuttgart 1971).

13 §§ 14, 27, 30, 46. Rudolf Steiner's scientific indications are scattered throughout his writings and lecture courses. The three main scientific courses are: *Geisteswissenschaftliche Impulse zur Entwickelung der Physik. Erster naturwissenschaftlicher Kurs* (Dornach 1964), English: *First Scientific Lecture Course. Light Course* (Forest Row 1978); *Zweiter naturwissenschaftlicher Kurs* (Dornach 1972); *Das Verhältnis der verschiedenen naturwissenschaftlichen Gebiete zur Astronomie. Dritter naturwissenschaftlicher Kurs* (Dornach 1926). Besides many lectures to doctors, Rudolf Steiner gave a course of lectures to farmers: *Geisteswissenschaftliche Grundlagen zum Gedeihen der Landwirtschaft* (Dornach 1975). English: *Agriculture* (London 1974).

14 § 14. George Adams: *Von dem Aetherischen Raume* (Stuttgart 1964); *Physical and Ethereal Spaces* (London 1978); *Strahlende Weltgestaitung* (Dornach 1965); *Die Pflanze in Raum und Gegenraum*, George Adams and Olive Whicher (Stuttgart 1979); *The Plant between Sun and Earth* (London 1980, forthcoming in French translation). Collected essays: *George Adams, Interpreter of Rudolf Steiner* (East Grinstead 1977). *Grundfragen der Naturwissenschaft* (Stuttgart 1979); *Nature Ever New* (Spring Valley, N.Y. 1979). Work material: *Universal Forces in Mechanics* (London 1977); *Universalkräfte in der Mechanik* (Dornach 1973); *A Letter from George Adams* (London 1978); *The Lemniscatory Ruled Surface in Space and Counterspace* (London 1979).

15 §§ 14, 27. Louis Locher-Ernst: *Raum und Gegenraum* (Dornach 1970); *Mathematik als Vorschule zur Geist-Erkenntnis* (Dornach 1973); *Geometrische Metamorphosen* (Dornach 1970); *Projektive Geometrie und die Grundlagen der euklidischen und polareuklidischen Geometrie* (Zürich 1940).

CHAPTER III

16 § 16. Projective Geometry. Besides the works of Adams and Locher, *Strahlende Weltgestaltung* and *Raum und Gegenraum* (see Notes 14 and 15), both of which have the advantage that they take their start from three-dimensional forms, there is an elementary introduction to Projective Geometry by Olive Whicher: *Projective Geometry, Creative Polarities in Space and Time* (London 1980), German edition, *Projektive Geometrie, Schöpferische Polaritäten in Raum und Zeit* (Stuttgart 1970). *Encounters with the Infinite* (Dornach 1971), German edition, *Begegnungen mit*

dem Unendlichen (Dornach 1970) by H. Keller v. Asten though not purely projective, provides very useful imaginative aids and exercises to the new geometry. Among other anthroposophical writers who have contributed to mathematics are: Elisabeth Vreede, Georg Unger, Alexander Strakosch, Ernst Bindel, Ernst Blümel, Ernst Schubert, Peter Gschwind, Lawrence Edwards, Arnold Bernhard, Ernst Kranich. Among the classical text-books to be recommended is the as yet still unsuperseded work of Theodor Reye: *Die Geometrie der Lage* (Leipzig 1923). There are a number of good elementary text-books, those for example by Cremona, L. N. G. Filon, and J. L. S. Hatton; Veblen and Young: *Projective Geometry* (Boston 1918); H. S. M. Coxeter: *Projective Geometry* (New York, London, Toronto 1964).

The lecture courses by Felix Klein (Berlin 1924–28) should be mentioned, which describe historical and biographical connections, and give information about the relationship of Projective Geometry to other fields of modern mathematics. Among the many more analytical works is J. A. Todd: *Projective and Analytical Geometry* (London 1947).

For more advanced studies we recommend C. Juel: *Vorlesungen über Projektive Geometrie mit besonderer Berücksichtigung der v. Staudtschen Imaginärtheorie* (Berlin 1932); E. Duporcq: *Premiers Principes de Géométrie Moderne* (Paris 1938). An interesting recent further development is Peter Gschwind: *Der lineare Komplex – eine überimaginäre Zahl* (Math.-Astron. Blätter, Neue Folge, Dornach 1977); *Methodische Grundlagen zu einer Projektiven Quantenphysik* (Dornach 1979).

17 §§ 16, 20, 21, 25, 27, 30, 31. *On geometrical terminology.* In relation to the ideas of the new geometry and about ethereal (negative-Euclidean) spaces and the forces at work in such spaces, we find it desirable to introduce certain termini technici, not merely to discard the old expressions, which are to be found in all the textbooks, but to provide mental pictures, which, while being equally exact, are not so one-sidedly bound to the fixed and finite concepts embodied in Euclidean geometry and the habitually pointwise conception of space.

Circle-curve, like conic-section, means any plane-curve, which can be transformed projectively into a circle; the emphasis here is rather on their *relatedness*, than on their determined form as sections of a cone. All real conic-sections have the same projective properties as the circle, and are in this sense various possible manifestations of the circle.

Line. The word "line" in this book means a straight line, as distinct from a curved line, which we call a "curve". *Line-Geometry* is the term normally used to mean a geometry which deals with the straight lines of space and the assemblages or manifolds – line-congruences, line-complexes – which thus arise. It conceives of the line not only in its pointwise aspect, but also as an assemblage of planes.

The Geometry in a Point. Rather than use the old terms "bundle" and "sheaf" when thinking of planes and lines running together in a point, which has a very spatial connotation, we change the quality of the idea. The two-dimensional manifold of planes and lines in a point (polar to the idea of points and lines in the plane) must awaken the thought that as well as the "extensive" entities as in Euclid's geometry, where quite intuitively we experience the parts as smaller than the whole, there must be entities in regard to which we need to learn to think the other way round. In projective space, it is equally true to think of whole planes and lines as being the parts of a point as it is to think in the more habitual way, that points and lines are parts of a plane.

The use of plane- and line- as well as point-co-ordinates in Analytical Geometry since Plücker fully confirms what is here intended. We are in fact making consistent use of the same principle which is adopted by Veblen and Young in saying that a line is *on* a point and not only that the point is on the line. Only we substitute the preposition *in.*

Ur-raum or Archetypal Space and the Principle of Polarity. Ur-raum or Archetypal Space is the name we give to the free projective space-potential of modern geometry, which is free of all metrical rigidity – a space ideally conceivable, the elements of which are points, lines and planes, whose relationships are given in the axioms of incidence, linear series and continuity. This may be briefly formulated: "As point is to plane, so is plane to point; as point is to line, so is plane to line." All the fundamental relationships evident to the imagination can be stated in pairs; figures or propositions, mutually related in this way, are called the *dual* of one another and the mental process of deriving the one from the other has been called *dualizing.* This is the so-called "Principle of Duality". The principle is better described as one of *polarity.* It is entirely consistent with the whole content and character of projective geometry (for three-dimensional space) to base it from the very outset on the perfectly balanced polarity of point and plane, without giving preference to either in the Axioms from which it starts. That this is not generally done, as for instance in the classical textbook of Veblen and Young, is due more to historical and traditional reasons. The word has already been adopted in the more specialized sense of polarity with respect to a conic, quadric, null-system, etc. We take the Principle of Polarity to be at the very foundation of projective geometry.

The same principle applies of course in a projective space of any number of dimensions, say n. The sub-space of $n-1$ dimensions, like the plane in three-dimensional space, is dual to the point, the space of $n-2$ dimensions to the line, and so on. It may be asked: Why do the authors attach so much importance to three-dimensional space; why not projective spaces of higher dimensionality, or again curvilinear spaces where the invariance of line and plane is abandoned? The hypothesis advanced in this book is indeed definite in this respect (which does not mean exclusive; other aspects too may ultimately prove important). It is, that the three-dimensional space, living in man's imagination and in his outer experience, is still of fundamental importance to the structure of the world and of its forces, but that the planar aspect is no less primary than the pointwise, the deeper underlying form being, in effect, projective.

As to the curvilinear spaces, these too – if any degree of continuity is to be preserved – depend on the primary entities of projective space, plane and line as well as point. The following passage from Felix Klein (Collected Papers, Vol. I, p. 482) is worth quoting in this respect. It is from his famous work, known as the *Erlanger Programm* (1872), which had so great an influence on the subsequent development of geometry. Klein draws attention to the fact that if the point is to be retained at all as a fundamental element of space, point-to-point transformation will always be "linear" in the infinitesimal domain. "In the infinitesimal the truths of Projective Geometry retain their validity. The projective aspect will therefore play a unique and outstanding role, no matter how wide may be the choice of transformation-groups in the science of geometrical manifolds."

Median Plane. The word "median" is used instead of "central" when describing a "planar centre", i.e. a plane playing the part of a middle (or central) organ in planar or counter-space, just as a point does in the familiar centric, Euclidean or pointwise space. The word *median* may also be used in relation to a line.

18   § 16. See Olive Whicher: *Projective Geometry*, Chapter III.

19   § 17. The necessary insight concerning the identity of the infinitely distant point in either direction of a sheaf of parallel lines dawned gradually. There should be mentioned here in particular Girard Desargues, who lived in the middle of the seventeenth century; then, about a hundred years later the distinguished Jugoslav philosopher R. J. Bošković; and at the beginning of the nineteenth century J. V. Poncelet, who also recognized the idea of the infinitely distant plane and the infinitely distant line.

20   § § 20, 34, 41. *The cone in a point, or intensive cone*, is a one-dimensional form, arising from a continuous succession of lines (generators) and tangent planes into one another in the two-dimensional geometry of a point (see Note 17). The naïve and one-sidedly pointwise, spatial consciousness sees the cone as a surface – if you like, as something two-dimensional in three-dimensional space. This is also justified, for after all one can draw two-dimensional pictures on the surface of a plaster cone. A clear distinction must, however, be made between this way of thinking and what is here meant, which is important for the correct understanding of the geometry of counterspace.

The cone considered as a surface lacks full plasticity; it is partially degenerate. As a surface it does have ∞^2 points, but only ∞^1 tangent planes. If one insists on regarding it as a surface, one would have to regard its polar form in Archetypal Space – the plane curve – as a surface. But the plane curve is degenerate in the polar opposite sense. It has ∞^2 tangent planes (for every one of the ∞^1 tangents of, say, a circle contains ∞^1 planes, all of which, without doubt, "touch" the circle) but only ∞^1 points. Purely ideally, and also in relation to the reality of Nature, it is an important experience to be able to think and feel the cone, not as an extensive, two-dimensional manifold of points, but as an intensive, one-dimensional form "within" a point.

21   § 22. Ellipsoid, paraboloid and hyperboloid.

22   § 22. In Steiner: *Goethe the Scientist*, the essay on *The Nature and Significance of Goethe's writings on Organic Morphology*.

23 § 23. Translation by George Adams in *Readings in Goethean Science* (Ed. Koepf; see Note 1).

24 § 23. Johann Wolfgang von Goethe: *Theory of Colours*, translated by Charles Lock Eastlake, with an introduction by Deane B. Judd, 1970 (M.I.T. Press, U.S.A. 1973); Rudolf Steiner: *Colour* (London 1979); M. Schindler and E. C. Merry: *Pure Colour* (London 1946); M. H. Wilson: *What is Colour?* (Clent 1949); *The Evolution of Light, Darkness and Colour* (The Golden Blade, London 1974); *Goethe's Concept of Darkness* (Journal for Anthroposophy, U.S.A., No. 24, 1976); *Das Dunkel als wirkende Macht* (tr. of *Goethe's Concept of Darkness*, Die Drei, Stuttgart, December 1977, p. 716); *Colour is Where You See It* (Die Farbe: International Congress Report 1965, p. 991); *Colour Dynamics – a painter experiments* (Palette, Basel, **37**, 1971, p. 33). See also the following articles by M. H. Wilson and R. W. Brocklebank: *The Complementary Hues of After-images* (J. Opt. Soc. Am., **45**, No. 4, 1955); *Goethe's Colour Experiments* (Physical Society Year Book, 1958, p. 12); *Two-Colour Projection Phenomena* (J. Phot. Sci., Royal Photographic Society, 8, No. 4, 1960); *"Land" Colour – a discussion* (J. Brit. I.R.E., **21**, No. 6, 1961); *Colour and Perception* (Contemporary Physics, 3, No. 2, 1961); *The Phenomenon of the Coloured Shadows* (Die Farbe: International Congress Report, 1961, p. 367); H. O. Proskauer: *Zum Studium von Goethes Farbenlehre* (Basel 1977); G. Ott and H. Proskauer: *Das Rätsel des Farbigen Schattens* (Basel 1979).

25 §§ 24, 26, 27, 31. *The Imaginary*. G. K. C. von Staudt (1798–1867) showed that the geometrical fact, answering to $\sqrt{-1}$ and to those spatial elements which in the usual analytical treatment receive imaginary co-ordinates, is in all cases a circling (elliptic) "involution" among the real elements. Take for example an ellipse and a point inside it. Analysis reveals a pair of imaginary tangents from the point to the curve. The geometrical fact is the pairing of real lines of the point, in polar-conjugate pairs with respect to the curve. (At the centre, it will be the pairing of conjugate diameters.) Such a pairing is known as an "involution". Followed continuously, it leads to a circling movement, the two lines always moving round in the same direction; the same form of movement can therefore be envisaged in either of two opposite directions. This is the geometrical reality underlying the idea of conjugate imaginary lines. The curve is as intimately related to this movement-form as it is to its real tangent lines. Von Staudt showed, moreover, that if the involution-movement in the one direction, say clockwise, represented the number $\sqrt{-1}$, then the same movement in the opposite direction represented $-\sqrt{-1}$. So in all cases, a pair of conjugate imaginary points, lines or planes signify the same circling involution in opposite directions. These "oriented involutions" accord with all the axioms expressing the mutual relations of points, lines and planes. The above example was in a real plane, but there are imaginary lines in space, containing no real plane and no real point. These are associated with spiralling families of lines (cf. Note 60).

When one has mastered the essentials, it is possible to make a mental picture of a purely imaginary curve or surface – say, of a sphere of given centre and radius $\sqrt{-1}$. There is no visible outline, but the form is, as it were, dynamically present, creating a definite field of relationships and forms of movement. The significance attributed in this book to the truths of projective geometry for the formative processes of Nature undoubtedly implies that the imaginary too plays a part.

The absolute imaginary circle in the plane at infinity determines the relations of right-angledness throughout space. Right-angledness is in the last resort a relation between families of parallel lines and of parallel planes, so that every line of the former is perpendicular to every plane of the latter. The lines lead to a point- and the planes to a line-at-infinity. The resulting one-to-one relation of points and lines at infinity proves to have all the properties of the pole-and-polar relation with respect to a conic. Only the conic is not real; the polar-conjugate relations give rise exclusively to circling movements and these invariably preserve the right angle. If, for example, the crystal shown in Fig. 49 belongs to the regular or cubical system – the one most intimately related to the measures and symmetries of Euclidean space as such – the 13 points and 13 lines of the harmonic pattern in the infinitely distant plane form 13 right-angled pairs, polar-conjugate with respect to the absolute imaginary circle.

Within Euclidean space, every sphere and every circle is determined by its intimate relation to this archetypal circle in the infinitely distant plane. A circle meets the line-at-infinity of its plane in the right-angled involution or pair of conjugate imaginary points determined there by the archetypal circle. They are the "circular points" of the plane. An imaginary line meeting the plane-at-infinity in a circular point is called a "circular ray".

In Projective Geometry, with the von Staudt theory of the imaginary (its association with the imaginary numbers is of course also significant) a perfectly clear and specific mental picture of an imaginary circle or conic can be formed, though it has no visible outline.

The absolute imaginary cone of negative space is necessary for the full determination of this space, as is the absolute imaginary circle for positive space. In the point-at-infinity of negative space there will be an imaginary cone, just as there is an imaginary circle in the plane at infinity of ordinary space (compare Note 20). A cone determines the mutual polar relationship of all the lines and planes of its point. In the imaginary cone obtained by this means every line will be perpendicular to its polar plane. We call it a spherical imaginary cone – one which determines a right-angled involution among the planes of every line and among the lines of every plane of the point which bears it (Filon, *Projective Geometry*). If the point-at-infinity of a negative space is located somewhere in positive space, as we shall be assuming, it will be simplest to conceive this imaginary cone as a direct perspective of the Euclidean absolute circle.

26 § 24. Concerning Grassmann see Felix Klein: *Entwickelung der Mathematik*, Vol. 1, 1926. Following the lines of thought of Projective Geometry, he proved that in spaces of any number of dimensions a polarity in relation to what we are here calling "expansion and contraction" must prevail. Grassmann: *Ausdehnungslehre* (Leipzig 1878).

27 § 26. *The Three Types of Measure. Derivation of Metric Spaces from Projective*. The differentiation of spaces – Euclidean, non-Euclidean and negative-Euclidean – from the more general projective space concerns the different types of measures there prevailing. In projective geometry there are no rigid measures; there are only functional relations finding expression in pure number – in anharmonic ratio, for example. "Measure" is essentially a rhythmic process, arising by the projection of a form – a line-of-points for example, or a line-of-planes – *into itself*. Suppose the point A to be projected into A_1. Regarded as a point of the untransformed line, A_1 is simultaneously transformed into A_2, likewise A_2 into A_3 and so on. Now there may be "invariant" points – points that remain ever untransformed, projecting back into themselves. Functionally, the constructions show that there are two such points; if there were more than two, *every* point would return into itself. Three possibilities arise. If the two points are real and distinct, the resulting measure has the exponential character of geometrical progression and is called hyperbolic; a suitable alternative name is *growth-measure*. If they are conjugate-imaginary, like the circular points in a line-at-infinity (cf. Note 25), it has the character of angular measure, leading to the circular or trigonometrical functions, and is called elliptic; more simply, it may be called *circling-measure*. The two invariant points may, however, also melt into a single real point; the outcome is then the stepwise measure (arithmetical progression) with which we measure lengths along a line in Euclidean space; this is called parabolic, or *step-measure*. Hyperbola, ellipse, parabola meeting the line-at-infinity in two distinct real points, in conjugate-imaginary points, and in a single real point counting doubly, reveal the three types of measure; hence the conventional names. The same three types of measure reappear in the line-of-planes and in the "plane pencil" or point-and-plane of lines.

In the above-mentioned rhythmic process the invariant elements act as *functional infinitudes*. A row of points – A, A_1, A_2, A_3, . . . – will perhaps tend towards one or other of the invariant points, but it will never reach it. The essential difference of growth-measure from step-measure is that the former has two distinct and the latter only one functional infinitude, since in the latter the two invariant elements have melted into one. The projective process can also be expressed algebraically; the difference emerges when this is done in the "canonical" or simplest form:

Step-measure, additive: $x' = x + a$; *Growth-measure, multiplicative:* $x' = kx$; where a, k are the parameters (other than 0 or ∞) by means of which the variable x is changed to x'. Asking the question: Is there any x which remains unchanged? – the resulting equation $x = x + a$ has only one solution, namely ∞; whilst $x = kx$ has two, namely ∞ and 0. The two play equivalent roles; they are indeed interchanged, without any change in the geometrical process, by substituting for x and x' their reciprocals. (∞ and 0 can also be replaced by any other distinct pair of real numbers; for instance by ± 1, when the growth-measure process finds expression in the "hyperbolic functions").

These things being understood, the metrical geometries can now be derived by planting in the projective space an invariant "Absolute", by relation to which the prevailing measures will arise. Four instances concern us here: Euclidean geometry and its negative or dual, and the two well-known non-Euclidean forms. In all of these, elliptic or circling-measure plays a part; they differ in respect of the two other types of measure. If the Absolute is an imaginary sphere (elliptic space) the latter do not arise at all; the sphere meets every line in a pair of conjugate imaginary points as well as planes. But if the Absolute is a finite *real* sphere (hyperbolic space), it meets a line, passing through the interior, in two distinct real points, and the relation to these two gives rise, in the line-of-points, to hyperbolic or growth-measure – hence the name of this type of space. Now if the Absolute becomes infinite, melting into a real "plane-at-infinity", the two distinct points in which it met the line merge into one. It is the famous point-at-infinity which in the new geometry is recognized as being one and the same at either end of the line. The line-of-points is now impressed with the familiar Euclidean step-measure, coming into play whether we step-out the length of a field or slide a yard-stick repeatedly from end to end. The same step-measure appears in perspective – in the same functional relation to the "vanishing point" – when in perspective constructions one creates a "vanishing scale", *échelle fuyante*; one sees then how the projective relation to the *single* point-at-infinity creates the Euclidean succession of steps of "equal length".

The expansion of the Absolute into a plane does not affect the circling measure prevailing in the line as a line-of-planes. The absolute sphere, when vanishing into the plane-at-infinity, still leaves behind the imaginary circle, and by relation to the latter every line-of-planes, passing through finite space, receives the angular or circling-measure with which we are familiar. Thus in Euclidean space any line passing through finite space has parabolic or step-measure as a line-of-points, and elliptic or circling-measure as a line-of-planes. For a line at infinity the opposite is true; as a line-of-points it has the circling-measure by which we always judge the angle between two directions, while as a line-of-planes, sending a sheaf of parallel planes into finite space, it has parabolic measure.

Precisely these relations are reversed in negative-Euclidean space, where the Absolute, instead of expanding into a plane, has shrunken into a point (see Note 32). Lines-at-infinity, namely the lines through the point U, have circling-measure as lines-of-planes by relation to the absolute imaginary cone, and step-measure as lines-of-points, with U as vanishing point of an *échelle fuyante*. All other lines in the surrounding space have circling-measure as lines-of-points and step-measure as lines-of-planes, with the plane through U acting as "infinitude within". This measure of a line-of-planes appears in its visually simplest form when the line is in the Euclidean plane-at-infinity ω and an initial plane is ω itself. The sequence of planes, coming in towards U, will then be parallel, and if the step from ω to the first succeeding plane – finite in negative space, infinite only in Euclidean – be chosen as the "unit distance", the succession of planes will in their "negative distances" inward from ω give the arithmetical series $0, 1, 2, 3, 4, \ldots$, which in Euclidean space, measured outward from U, appears as the reciprocal – as the harmonic series $\infty, 1, \frac{1}{2}, \frac{1}{3}, \frac{1}{4}, \ldots$

28 § 26. See the elementary exercises illustrating the influence on measure of the infinitely distant elements in Olive Whicher: *Projective Geometry*.

29 § 26. Geometrically, the elementary crystal lattice is a harmonic network, formed from an archetypal harmonic pattern in the plane-at-infinity; in Fig. 49 this is replaced by a horizontal plane through finite space, to make the process more visible. From the points and lines of the archetypal pattern, lines and planes ray out into three-dimensional space to form the crystal lattice, of which a cubical and an octahedral cell are here seen (the one within the other as in Fig. 35); the cubical cell is already subdivided into eight smaller ones. The weaving process thus begun goes on potentially for ever, filling the whole of space.

The conventional projection of the crystal faces on to a sphere with the crystal at the centre, leading to the Millerian indices, etc., represents, in effect, the pattern in the infinite sphere, the plane-at-infinity, from which the crystal derives. The same type of harmonic network is fundamental to the constructions of perspective.

30 § 26. D'Arcy Thomson: *On Growth and Form* (Cambridge 1942). See also Note 64.

CHAPTER IV

31 §27. Ernst Lehrs: *Man or Matter* (London 1951), *Mensch und Materie* (Frankfurt 1953).

32 § 28. A. N. Whitehead, *Universal Algebra* (Cambridge 1898), § 239.

In Adams/Whicher: *The Living Plant* (see introduction to the present book), this important mathematical conception was formulated as follows in the notes: *Euclidean, Non-Euclidean and Negative Euclidean Geometries: Continuous Transition*. Modern geometry provides an interesting line of thought, leading to the "negative Euclidean" type of space by a continuous metamorphosis. We have mentioned that non-Euclidean spaces as well as Euclidean are derivable from projective geometry. Only the "Absolute" must then be imagined, not as a plane but as a finite sphere, with a definite, real or imaginary radius (more generally, as a quadric surface). Two types of non-Euclidean space result, according to whether the sphere is real or imaginary. The ideal transition from non-Euclidean space to Euclidean can then be accomplished by letting the radius of the "absolute sphere", originally finite, grow beyond all finite limits. The real or imaginary sphere is then turned into a real plane (doubly covered) – a plane that bears, like a trace or phantom of the sphere from which it comes, the absolute imaginary circle. Moreover we can pass from the finite *real* sphere via the infinite plane to the finite *imaginary* one, so that the two non-Euclidean geometries appear on either side of the Euclidean, with the latter as the transition-state between them. Euclidean geometry is then called "parabolic" and the two non-Euclidean geometries "hyperbolic" and "elliptic" respectively.

From the aspect of pure thought there is again a satisfying harmony in these relations. Yet to complete the train of thought one link is missing still. Both arithmetically and geometrically there is a second way to make the transition from a real to an imaginary sphere. Instead of letting the sphere expand into a plane, we let it contract into a point and thence grow outward as an imaginary sphere. The finite real value of the radius becomes zero; it grows again as an imaginary number. If we conceive this transition – the necessary ideal complement of the former one – we are again led to the negative-Euclidean (or negative parabolic) type of space with a "point-at-infinity" instead of a "plane-at-infinity" for Absolute. It is a *hyperbolic planar space*, polar with respect to the absolute sphere to the *hyperbolic pointwise space* inside the latter. Therefore in hyperbolic geometry the real Absolute gives rise to two non-Euclidean spaces – a pointwise one, like a more or less distorted modification of ordinary space, inside the absolute sphere, and a planar one outside it. Professor A. N. Whitehead (*Universal Algebra*, § 202) calls the former one the Space, the latter the Anti-space. If we expand the absolute sphere into the infinite, the Space becomes parabolic or Euclidean; the Anti-space melts into nothing. If on the other hand we contract the sphere into the central point, it is the pointwise Space inside which vanishes. The planar Anti-space becomes the negative-Euclidean type of space we have described; its Absolute is then the point-at-infinity into which the sphere has contracted, bearing the spherical cone like an echo of the sphere.

In the elliptic form of non-Euclidean geometry with its imaginary Absolute there is no sundering of Space from Anti-space; there is only the dual, pointwise and planar aspect of the one undivided Space. Here therefore the principle of duality holds no less perfect sway than in projective geometry pure and simple. The nineteenth century mathematician W. K. Clifford (*Lectures and Essays*, Vol. 1, p. 322) sought escape from the rather "tantalizing" position as we have called it – the breakdown of the pure polarity of space in metrical geometry – by supposing that cosmic space was after all elliptic and not parabolic, but with so large an imaginary Absolute that with our spatial measurements we had not yet been able to detect the difference. "Upon this supposition," he writes, "the whole of geometry is far more complete and interesting; the principle of duality, instead of half breaking down over metrical relations, applies to all propositions without exception. In fact I do not mind confessing that I personally have found relief from the dreary infinities of homaloidal (i.e. Euclidean) space in the consoling hope that, after all, this other may be the true state of things."

The problem is solved in quite another way if we frankly recognize that physical space is Euclidean, but that this space is only the one pole in the living processes of the spatial Universe, the other being the ethereal or negative-Euclidean. As to the "dreary infinities", they appear in quite another aspect when we find in them the archetypal source of the ethereal powers which renew and replenish life.

Occasional references to the negative-Euclidean type of space will be found in the earlier literature. The two-dimensional, plane geometry of this type is briefly dealt with by Felix Klein in his lectures on Non-Euclidean Geometry (Springer, Berlin 1928, p. 182). In a paper in the "Proceedings of the Edinburgh Mathematical Society", Vol. 28, 1910, Professor

D. M. Y. Sommerville enumerated a large number of conceivable geometries of three dimensional space on the basis of the Cayley-Klein theory; among them was the polar opposite to that of Euclid – the geometry which we are calling "negative-Euclidean". Such possibilities were mentioned, only to be left aside.

33 § 28. W. K. Clifford: *Lectures and Essays* (London 1918), Vol. 1, p. 322.

34 § 30. The idea of "force" in physics and mechanics has undergone many developments since Newton's time. In relation to work or energy it appears as the differential coefficient of kinetic or potential energy with respect to distance. In thermodynamics this has opened out the way to a wider concept; force and related magnitudes appear as "intensity factors" in the equations of energetics. Fields of potential force are an essential feature in electro-magnetic theory. Increasingly, the "forces", with all the mathematical properties which they reveal, are recognized as belonging integrally to the dynamic structure of the phenomena; in this the element of time as well as the forming of manifold spaces is involved.

In working out the "dual" concept – that of planar forces in negative-Euclidean spaces – we purposely adopt a rather simple and old-fashioned standpoint, taking our start from the elementary ideas of classical mechanics. Ultimately there is little doubt that the two polar-opposite realms, thus contemplated side by side, will prove to be more closely interwoven. The interplay of ponderomotive force with heat and other forms of energy will in all probability appear in a new light when the "planar" dynamic realm is recognized in its real connection with the pointwise. There are, however, phenomena predominantly if not purely mechanical in nature (approximations to the "conservative systems" of classical theory), for which Newtonian mechanics may be said to express – to use Goethe's term – the Ur-phenomena. It is helpful at the present stage to confront this realm with its precise polar opposite, in order then to work from both directions toward a middle realm more realistic than either of these ideal thought-forms.

35 § 30. Rudolf Steiner: *Die Bedeutung der Anthroposophie im Geistesleben der Gegenwart*, lecture of 9 September 1922 (Dornach 1957). See also Rudolf Steiner and Ita Wegman: *Grundlegendes für eine Erweiterung der Heilkunst nach Geisteswissenschaftlichen Erkenntnissen* (Dornach 1925). English: *Fundamentals of Therapy* (London 1925).

36 § 31. The word *intensity* applied to the planar analogue of mass is consistent with the use which has already been made of this word in certain purely geometrical calculi (Grassmann, A. N. Whitehead), the first of which was the "Barycentric Calculus" of Moebius (1827). They take their start from the far-reaching harmony between the projective structure of space and the experimentally determined laws of statics. Our suggestion is, in effect, that the same harmony also extends to forces of a "dual" kind, and that the geometrical and well-proportioned beauty of living forms is due to this.

37 § 31, 32. The determination of "planes of levity" is, in principle, a simple exercise in projective geometry – a kind of "barycentric calculus" in negative. To take the simplest example, two planes of equal levitational intensity have a common line, say l, which with the point-at-infinity of the space gives a third plane, lU. The harmonic fourth plane which lU determines with respect to the given pair is then the plane of levity. If the planes are of different intensity, the ratio of their intensities, multiplied by -1, gives an anharmonic ratio which fixes the plane of levity. This is the precise analogue of the law of moments; the latter need only be re-stated in projective terms by the inclusion of the point-at-infinity as a fourth point along the line of the lever.

We think of the principle of "levity" or "active buoyancy" in contrast to hydrostatic buoyancy as explained by the principle of Archimedes, where a heavy body is buoyed up by the surrounding liquid and will even float to the surface if it is specifically lighter. In this case the resultant upward thrust is conceived not as an active principle of lightness, but, so to speak, as the result of a subtraction sum, still in the realm of weights and pressures. It is in this sense a passive and rather negative kind of buoyancy. The planar force we are here supposing will be as active and positive in the realm of "lightness" as is the force of gravity in the realm of weight. It will no longer be lightness in the sense of very little weight, but positive and active lightness; we call this "levity".

The possibility of developing a theory of levitational or ethereal forces along these lines will not be doubted. The geometrical polarity of spaces positive and negative is so complete that every competent mathematician who entertains the thought will see that it can be done. Some

212

of the most important thought-forms of physical statics and dynamics are already imbued to no small extent with the inherently polar forms of projective geometry; one-sidedly physical as they are, it is as though they were asking for their polar counterpart.

A wide range of facts and phenomena concerning the physical and ethereal aspects of the Earth-planet will be found in G. Wachsmuth's works: *Erde und Mensch*, 1945, and *Die Entwickelung der Erde*, 1950. *Etheric Formative Forces in Cosmos, Earth and Man* (London 1932) (referring to §32). If, in addition to negative gravity or levity, one attributes to the planar entities in negative space a dynamic property analogous to the "inertia" or momentum of material particles in Euclidean space, and a corresponding law relating force and acceleration, an "ethereal dynamics" can of course be worked out. The concept leads to interesting conclusions at the very outset. In physical mechanics, broadly speaking, the tendency of "inertia" – if there is movement at all – is centrifugal, while that of gravity is centripetal. In the ethereal the relations are reversed. The simplest picture of an ethereal pendulum, swinging in simple harmonic motion, is of a planar entity displaced from its equilibrium position in the physical plane-at-infinity ω and swinging in up to a certain distance towards U, the infinitude within, then swinging back to ω and through, approaching U from the other end of space, and so on. At the end-planes the swinging plane has a maximum "ethereal potential energy". A corresponding picture, ethereally equivalent, only appearing less symmetrical in the perspectives of physical space, will arise if the equilibrium position is a plane other than ω. It would be interesting to study from this aspect the rhythmic opening and closing movements of leaves and flowers.

If the dynamic conditions are more complex – as in the type of physical pendulum forming the Lissajous figures – a planar entity will in general follow a spatial path-curve, of which it is the osculating plane. It will at every moment turn about the tangent line to the path-curve, the differential conditions of instantaneous movement being dual but otherwise equivalent to those of Newtonian mechanics. Moreover as the path-curve of a material particle often becomes a plane curve, that of a planar entity may become a cone.

An interesting possibility arises for the rotation of the Earth on its axis, supposing that the ethereal planet shares the diurnal movement. Just as the physical Earth through its rotational momentum bulges out at the equator, the ethereal will tend to draw in centripetally, bringing about from either side a greater mutual penetration of the physical and ethereal in the tropics and a drawing asunder at the poles.

38 §§ 31, 43. The word *spheroid* is here used in a wider sense, to include ellipsoid, elliptic paraboloid and two-branched hyperboloid – any quadric surface with "positive curvature". In the descriptions of negative space, differences of name which apply only to the relation of the form to the physical plane-at-infinity are often an unnecessary ballast.

39 § 31. Compare the last paragraph in Note 27.

40 §§ 32, 36. H. W. Turnbull, *Mathematics in the Larger Context*, Research, Vol. 3, No. 5, 1950.

41 § 33. The fundamental idea of deriving different types of space from projective was set forth by Felix Klein in his "Erlanger Programm" quoted in Note 17. Within the group of all projective transformations one envisages the sub-group of all those which leave a certain fundamental form – such as an absolute quadric, or a plane or point at infinity – invariant. The invariant form can also grow more complicated, but a limit is quickly reached, since every added determination restricts the freedom of transformation. Therefore in recognizing that a living organism contains more than one potential point-at-infinity, it will not primarily be a question of investigating the projective movements which are still possible when several are fixed, but rather of conceiving how the same planar entities may be successively and even partially related to the one and to the other. And this again, both as to form and force (cf. Note 42), becomes a problem in pure projective geometry.

Yet the theory of transformation-groups, even with fixed invariant elements, has an important bearing. In the most general type of continuous group of "collineations" there is an invariant tetrahedron, two or all of the points (and therefore planes) of which may form conjugate-imaginary pairs. Some of the path-curves and surfaces which then arise (W-curves and -surfaces of Klein and Lie) have an obvious relation to living forms. (Cf. Note 60.) Such likeness generally comes about when one of the invariant elements – a line or plane, for example, of the tetrahedron – is in the Euclidean infinitude; it gains a new significance when it is seen that another element, opposite to this, is at the same time in the infinitude of the

213

ethereal space of the living organism. It is also important that in the spatial pictures which arise when one imagines these transformation-groups, one is obliged to operate not only with point, line and plane as such, but with their syntheses: point-line, line-plane, plane-point and point-line-plane, meaning in each case the entity consisting of two or more of the three basic elements, incident in one another. (Along a spatial path-curve, for example, point, tangent line and osculating plane move as an inseparable triad. See the many elegant descriptions and pictures in Sophus Lie and Scheffers: *Geometrie der Berührungs-Transformationen*.) Projective geometry here leads into forms of thought bearing directly on that "middle realm", mentioned in Note 34 where the secret is still to be discovered of the real interplay of the pointwise and the planar, the physical and the ethereal in Nature.

42 § 33. *Projective theory of forces. Geometry and Mechanics*. Among the elementary truths of mechanics is the fact that three forces can only be in equilibrium if their lines of action meet in a point *and* in a plane. The condition – necessary though not sufficient – is self-dual and projectively invariant. A rather less trivial example from the statics of rigid bodies: If the lines of action of three forces are skew, a fourth force can be found to balance them; its line of action always belongs to the same regulus (system of hyperboloid-generators), which the three determine, once again, by purely projective laws (Routh, *Analytical Statics*, 2nd Edition, 1909, Vol. I, § 316). The close connection of mechanics and projective space is also shown by the importance in the former of the "null-system" and by the derivation of reciprocal force-diagrams (Maxwell, Cremona). Once the importance of a theory of forces in purely projective space, and of forces planar as well as pointwise, has been recognized, much that is relevant will be found in the existing literature, e.g., in the work of F. Lindemann (*Math. Annalen*, 1873), A. N. Whitehead (*Universal Algebra*, 1898, Book V), Sir Robert Ball, Felix Klein and E. Study. See also the more recent work indicated in a preliminary note by J. M. T. Vidal, "Fondements d'une mécanique projective", *Comptes rendus*, T. 252, 1951, p. 2397.

43 § 33. Kranich: *Die Formsprache der Pflanze* (Stuttgart 1979). Blattman: *Die Sonne – Gestirn und Gottheit* (Stuttgart 1972). See also Note 45.

44 § 34. F. H. Julius: *Metamorphose: een Sluitel tot Begrip van Plantengroei en Menschenleven* (Den Haag 1948), German: *Metamorphose, Ein Schlüssel zum Verständnis von Pflanzenwuchs und Menschenleben* (Stuttgart 1969).

45 § 36. Rudolf Steiner repeatedly described what science must one day find the true nature of the Sun to be. See for example: the Second Science Course *Geisteswissenschaftliche Impulse zur Entwickelung der Physik. Zweiter naturwissenschaftlicher Kurs* (Dornach 1972), Lectures 1 and 14, and the Third Science Course *Das Verhältnis der verschiedenen Naturwissenschaftlichen Gebiete zur Astronomie. Dritter naturwissenschaftlicher Kurs* (Dornach 1926), Lecture 18.

46 § 37. Even the so-called "velocity of light" in its physically spatial effects is seen in quite another aspect since Einstein's and Minkowski's work at the beginning of the present century. Moreover the actual measurements of this "velocity" depend on the intervention of darkness in a very concrete form – for example, on the periodic extinction and release of flashes of light by the rotating wheel in Fizeau's experiment. See the detailed reflections on this subject by Dr. E. Lehrs in Chapter XVII of *Man or Matter* (London 1951, reprint planned).

47 § 37. For "light" and "dark", the German language as used by Goethe includes a greater variety of expressions, with the attributes *hell* and *dunkel* as well as the nouns *Licht* and *Finsternis* and the adjectives derived therefrom. In *Man or Matter*, Chapter XVI, Dr. E. Lehrs proposes that the words *Light* and *Dark* should be used for the primary, ideal polarity, reserving the expressions *lightness* and *darkness* for visible effects.

48 § 37. Although in quality of thought the undulatory theory with its picture of a wave-front surface would appear less remote from the ethereal than the corpuscular, wave-length in its accepted meaning is none the less a typically Euclidean concept. Marking in thought successive equal phases along a "ray", or normal to the wave-front, one has a sequence of points or tangent planes, in "parabolic" or "step-measure" exactly as in a crystal lattice, or in the harmonic networks mentioned in Note 29. The phenomena on which the wave-theory is based are due to the interaction of light with matter in its purely physical, inorganic aspect. The crystal lattice – shown by X-ray analysis to be far more prevalent in the solid and even semi-solid phase of matter than used to be supposed – is indeed the "Ur-phenomenon" of purely physical space. Experience and thought agree in this. The harmonic network with its additive (parabolic) measures in all directions is the first form one comes to when working out

the consequences of a unique plane-at-infinity in an otherwise projective space. Matter thrown out of the realm of life is, as it were, precipitated into the space determined by the Euclidean plane-at-infinity. As such, it tends naturally to the crystal-lattice form. Is it not equally natural that the light itself, when entering this realm, should become rhythmicized in the same kind of spatial measures? Seen from this point of view, the ethereal aspect of light (using the word "ethereal", needless to say, in the sense here defined and not with reference to the nineteenth century "luminiferous ether") need not be inconsistent with facts established in the inorganic realm.

49 § 37. Rudolf Steiner: *Geisteswissenschaftliche Impulse zur Entwickelung der Physik. Zweiter naturwissenschaftlicher Kurs,* Lecture 11 (Dornach 1972).

The word *intensity* applied to the planar analogue of mass is consistent with the use which has already been made of this word in certain purely geometrical calculi (Grassmann, A. N. Whitehead), the first of which was the "Barycentric Calculus" of Moebius (1827). They take their start from the far-reaching harmony between the projective structure of space and the experimentally determined laws of statics. Our suggestion is, in effect, that the same harmony also extends to forces of a "dual" kind, and that such concepts must be applied in the whole of physics.

CHAPTER V

50 § 38. There has been a great development of the new style in architecture since this book was first published. Throughout Europe and in many parts of the world, the first attempts have been made to follow Rudolf Steiner's indications in the so-called "Goetheanum Architecture", in the building of schools, clinics, anthroposophical centres and private dwellings. When describing the Goetheanum forms, Rudolf Steiner tried again and again to awaken an understanding of true earthly and cosmic polarities. In the introduction to the publication of one of his lecture courses, entitled *Ways to a New Style in Architecture* (London 1927), *Wege zu einem Neuen Baustil* (Dornach 1927), Marie Steiner includes the words of a Californian architect, with reference to the two cupolas of the First Goetheanum: "The man who has solved this problem is a mathematical genius of the highest order. He is a master of mathematics, a master of our science: from him we architects have to learn. The man who built this has conquered the heights because he is a master of the depths."

51 § 38. Theodor Schwenk: *Sensitive Chaos* (London 1976); *Das Sensible Chaos* (Stuttgart 1976); *Le Chaos Sensible* (Paris 1963). See also paragraph 46. Schwenk was until quite recently the leader of the research institute in the Black Forest, founded together with George Adams, Dr. Alexander Leroi, Dr. Georg Unger and others, with the support of Dr. Hanns Voith of Heidenheim. Wolfram Schwenk now leads the Institute: Institut für Strömungswissenschaften, Herrischried, D–7881, West Germany.

52 § 39. A similar idea is suggested by G. Grohmann in: *The Plant* (London 1974) translation of Vol. 1 of *Die Pflanze* (Stuttgart 1959.). French: *La Plante* (Triades, Paris 1978).

53 § 40. With the concept of a "functional infinitude within" of a negative space, we take hold of one of Rudolf Steiner's important statements out of Spiritual Science. He frequently insists on the need for natural science to discover the secret that in the seed, earthly matter must at first become completely formless, like a chaos, in order to receive from the universe the form-giving influences for the newly arising individual. See *Das Verhältnis der verschiedenen naturwissenschaftlichen Gebiete zur Astronomie. Dritter naturwissenschaftlicher Kurs* Lecture 4 (Dornach 1926). See also E. Vreede, *Anthroposophie und Astronomie* Chapter 32 (Freiburg 1954).

54 § 40. In connection with the two-way stream of time, see Rudolf Steiner's lecture of 22 March 1922. Phenomena, which indicate that the habitual cause and effect manner of thinking of natural science is insufficient, have been investigated by other researchers. See, for example, J. W. Dunne: *An Experiment with Time* (London 1929) and the dissertations of C. G. Jung: *Synchronizität als ein Prinzip akausaler Zusammenhänge* ("Naturerklärung und Psyche", Zürich 1952). English: *Synchronicity: An Acausal Connecting Principle* (London 1972).

55 § 40. G. Adams: *Universal Forces in Mechanics,* see Note 14.

56 § 41. This is a logarithmic-spiral cone, in an unending sequence of folds winding in towards the vertical axis and out towards the horizontal plane. The cone is here conceived as a continuous one-dimensional form of lines and planes (generators and tangent planes) in the two-dimensional geometry of the point which bears it (cf. Note 20).

57 § 44. Goethe repeatedly suggests that from the root towards the flower a refinement of substance is taking place; this he associates with the influence of light and air and with the progressive elaboration and delicacy of the forms. (See for example *Metamorphosis of Plants*, §§ 12, 13, 30, 39, 64, 120.) Though this may seem problematical in terms of the now prevailing chemical ideas, it may gain quite another significance when taken in conjunction with the physical-ethereal polarity.

58 § 44. Goethe: *Über die Spiraltendenz der Vegetation*, 1831. Kürschner edition, Vol. I, pp. 217–238.

59 § 44. The ancient picture of the *Caduceus* or Mercury's staff – a vertical rod entwined with twigs or serpents – is here used to typify the interplay of vertical and spiral principles in the green leaf-bearing stem. This is the characteristic middle region of plant growth, intermediate between the ultimate polarities of root and flower. A realm of rhythmic repetition and renewal, it represents what the old alchemists would have described as the "mercurial" principle, intermediate between "salt" and "sulphur".

60 § 44. Among the forms in which a pair of skew lines – such as a vertical line and the infinite horizon – play an important part are the null-system (linear complex) and the concentric family of hyperboloid-reguli forming together the "elliptic linear line-congruence", the lines of which bear the points and planes of an imaginary line in space, as shown by von Staudt (cf. Note 25). Through its relation to the imaginary, this spiralling form of lines is one of the most fundamental elements in the structure of space altogether.

There are important spiralling transformations in projective space for which a pair of skew lines are the only *real* lines of the invariant tetrahedron (Note 41). If for example a line-at-infinity with its circular points helps form the tetrahedron, path-curves or lines of flow arise which are quite evidently akin to spiral forms in Nature, notably in the animal kingdom. The relation to the plant is on a deeper level. A spiral *ordering*, unquestionably mathematical in character, is shown in the phyllotaxis. With the exception of the climbing plants, however, outstandingly spiral forms seem rather to occur in abnormal growths, such for example as the Valerian, of which a picture, published by Goethe, is reproduced in Dr. Arber's edition of the *Metamorphosis*. This would suggest that the "law-giving power in the midst" transforms a latent tendency into a far more delicate form of expression.

61 § 45. In a lecture cycle given in The Hague to university graduates from 7–12 April 1922: *Die Bedeutung der Anthroposophie im Geistesleben der Gegenwart* (Dornach 1957), Rudolf Steiner speaks a great deal about the hollow, or negative spaces, into which "forces in surfaces (planes) come in from all sides of the universe to mould from the outside the living forms". He describes the need to complement the one-sidedly spatial thought-forms (spatial according to the cartesian system) by thinking also of the dimensions as though negated, thus reaching to the conception of a point, which is like no ordinary point, but a dimensionless realm, "a point which leads us spiritually to something most high, provided we can comprehend it, not as an empty point, but one that is filled".

62 §§ 45, 49, 52. *Lemniscatory or Cassini Spaces*. Plates X, XI (top left), XVI and XIX. These forms arise when the exponential rates of simultaneous inward and outward increase in the two families of spheres or circles are freely chosen. Let U, O respectively denote the upper and lower or "light" and "dark" focus. Then with respect to a time-parameter t, the (Euclidean) radii r_U, r_O will be respectively proportional to e^{-nt} and e^{mt}, where m, n are positive real numbers. If the ratio $m:n$ is rational, the resulting curves will be algebraic. With these assumptions we may take (m, n) to be integers, mutually prime. Their values are $(1, 1)$ in the ordinary Cassini curves, Plate X (bottom right); $(2, 1)$ in Plate XI (top left); $(16, 1)$ in Plate XVI and $(3, 2)$ in Plate XIX.

It is convenient to write the factors of proportionality in the forms me^{nc} and ne^{mc} for r_U, r_O respectively, where c is a parameter distinguishing the curves or surfaces of one family; units will presently be chosen so as to assign the value $c=0$ to the lemniscate. We therefore write:

$$r_U = me^{n(c-t)}, \qquad\qquad r_O = ne^{m(c+t)}, \qquad\qquad \text{(i)}$$

and by elimination of t we obtain the constant product:

$$r_U{}^m r_O{}^n = m^m n^n e^{2mnc}. \qquad\qquad \text{(ii)}$$

A simple geometrical reflection shows that a node or double point will arise along the axis OU if for the circles touching at this point the rates of increase and decrease, seen in the aspect of Euclidean pointwise space, are equal, in other words if $dr_U/dt=-dr_O/dt$. This is consistent with equation (ii) only if $r_U:r_O=m:n$, confirming what was indicated in § 45: the distances of the node from U and O are in inverse proportion to the exponential factors n and m. If we choose units so that $OU=m+n$, the numbers m and n will represent the distances of the two foci, U and O respectively, from the node. $c=0$ will then give the lemniscate the value $t=0$ for the time-parameter corresponding to the node itself. Taking the origin at the node and substituting the parameter a for c by writing $e^{2mnc}=a^{m+n}$, the equations of the (m, n) family will then be:

$$\left\{ x^2 + (y-m)^2 \right\}^m \left\{ x^2 + (y+n)^2 \right\}^n = m^{2m} n^{2n} a^{2(m+n)}, \tag{iii}$$

revealing that the (pointwise) order of the curve is $2(m+n)$ and giving the familiar form for the normal Cassini curves if $m=n=1$.

Equation (iii) confirms that for $a=1$ there is a node at the origin; moreover the curve at this point approximates to $x^2-y^2=0$, showing that the nodal branches cross at right angles. The node is not a point of inflexion unless $m=n$; the inflexions lie along the larger loop. This loop becomes more and more nearly circular, as in Plate XVI, when the ratio $m:n$ grows large. The parameter a has a simple geometrical meaning. For the pair of U- and O-circles whose radii are ma, na, the rates of decrease and increase, seen in Euclidean space, are momentarily equal as at the node of the lemniscate.

The eccentric Cassini space has (top right in Plate XI) been formed by the interpenetration of concentric spheres growing outward from the "dark" focus O (shown in green) and spheres coming inward towards the "light" focus U (shown by the star) also in geometrical progression, but with the focus placed eccentrically. The diameter through the eccentric focus is vertical. If C is the Euclidean centre of any one of these spheres, the ratio of CU to the radius measures the degree of eccentricity. The inward or outward growth of the circles being a similarity-transformation (homology between U and the plane-at-infinity), this eccentricity remains the same for all the spheres.

Geometrically, the resulting forms depend on three variable factors: (1) the eccentricity of the upper spheres, assuming the lower ones to remain concentric; (2) the angle between the line OU and the direction of eccentricity CU; and (3) the ratio $m:n$ defined as above. In Plate XI the eccentricity is $2:3$, the angle OUC is $126°$, and $m:n=1$ as in the ordinary Cassini curves.

The eccentricity in the vertical dimension obviously means that the ethereal space is modified by relation to the gravity- and levity-field of the Earth-planet as a whole. In form and gesture this would often seem to be so in the less symmetrical type of flower, as in the Labiate pictured in Plate XVIII. The flower forms its own individual relation to the Earth below and the celestial periphery above, in relative detachment from the plant on which it grows. It is an interesting fact that precisely the Labiates with their high degree of symmetry in leaf and stem produce this type of flower.

63 § 45. A clear picture of the transition from the one to the other branch of the double oval can be formed on the basis of the von Staudt conception of the imaginary, mentioned in Note 25.

64 § 45. Expansion of spheres, circles or other forms in geometrical progression from a given centre, say O, towards the plane-at-infinity ω (or their contraction from ω towards O) provides the simplest visual picture of the multiplicative process $x'=kx$, which by repeated application: $x''=kx'$, $x'''=kx''$, etc., leads to the potentizing series: x, kx, k^2x, k^3x, ... Projectively, this is a "homology" between O and ω, with k or k^{-1} as anharmonic ratio. The process is equally determined from the periphery inward and from the centre outward – self-dual, perfectly balanced between the two. If therefore O is the point-at-infinity of an ethereal space, the geometrical progression between O and ω expresses in the most natural way the mutual relation.

The logarithmic spiral as a phenomenon in Nature has claimed the attention of the artist and the scientist through the centuries. A great amount of fascinating material may be found in the following well-illustrated books: *Design in Nature* by J. Bell Pettigrew (London, New York, Bombay and Calcutta 1908); *The Curves of Life* by Theodore Cook (London 1914); *On Growth and Form* by D'Arcy Wentworth Thompson (Cambridge 1942); *Pflanzengeometrie* by Werner Schüpbach (Bern 1944); *The Mystic Spiral* by Jill Purce (London 1975). Although these books treat the spiral and other curves in the old way, according to Euclidean geometrical concepts, they bring illustrative material in abundance, the study of which, when enriched by the freer ideas about form provided by the new geometry, awakens a deeper understanding of the very forms of the living body, suggesting the intimate interweaving of physical and ethereal formative processes, which must have been at work in the creation of those forms.

George Adams, towards the end of his life, in his researches in connection with water together with Theodor Schwenk (see Note 51), developed spiralling forms and surfaces of a higher order than spirals, based on the principle of path-curves and their associate invariant surfaces (see Note 41). One of his students, Lawrence Edwards, Edinburgh, continued to work at these surfaces after Adams' death and discovered their relationship to forms in nature, such as pine-cones, buds and animal organs, and also the vortex, which have hitherto been described as spirals. The path-curve-surfaces developed by Adams are also based on growth-measure and are forms between the Euclidean infinitude and the innermost verticon. Edwards' work on the plant forms is a significant further development of the work described in this book. He has published essays in the *Mathematical-Physical Correspondence*, edited by Stephen Eberhart, Dept. of Mathematics, University of Montana, U.S.A. (See in particular the essay entitled *Ethereal and Physical Spaces in Flower and Fruit Forms*.)

CHAPTER VII

65   § 48. *The Archetypal Plant (Ur-pflanze)* and the leaf as an archetypal form. In the essay entitled *Origin of the Theory of Metamorphosis*, in his introductions to Goethe's scientific work (see Note 1), Rudolf Steiner writes:
On May 17th 1787, Goethe wrote this thought to Herder. On the archetypal plant itself, Steiner contributes the following: The living entity is a whole enclosed within itself, which produces its states of existence through its own nature. But in the juxtaposition of the members and in the chronological succession of the states of existence of a living entity, there is a reciprocal relationship which does not come to manifestation through a determinative influence of the sensible characteristics of the members, through the mechanical-causal determination of the later by the earlier, but is controlled by a higher Principle, belonging above the members and the states of existence. It is inherent in the nature of the whole that a definite state is fixed as the first and another as the last; and the succession of the intervening states is also determined within the idea of the whole. The preceding is dependent upon the succeeding and vice versa. In short, in the living organism the evolution of one out of the other, the transition of states one into another, is no ready-made, finished existence of the single entity, but a constant *becoming*. In the plant this determination of each single organ by the whole comes to manifestation to the extent that all organs are built upon the same fundamental model.

66   § 48. Goethe: *The Metamorphosis of Plants*, § 15.

CHAPTER VIII

67   § 51. Goethe: *Metamorphosis of Plants*, § 106.

68   § 51. Goethe: *Theory of Colours*, Introduction.

69   § 52. Goethe: *Faust* (part two) last scene.

70   § 54. Goethe: *Theory of Colours*, § 919.

71   § 54. Rudolf Steiner: *Das Wesen der Farben* (Dornach 1973); *Colour* (London 1979).

72   § 57. The process of seed-formation is here described in continuity with the whole morphological picture, evoked by the entire plant through all the phases of its growth. This is in harmony with Goethe; as he himself says (*Metamorphosis*, § 63), he was glad when he could bring nearer to each other the ideas of growth and generation. "The healthy course of metamorphosis implies that the living plant, rooted in the soil and striving upward stage by stage into light and air, should by its own enhancement at long last bring forth the culminating seed, to be detached and scattered." (Freely translated from a passage where he described his own relation to Schelver's rather unorthodox opinions. From the article "Verstäubung, Verdunstung, Vertropfung", 1820. Kürschner's edition, Vol. I, p. 160.)

It should scarcely be necessary to point out that this description does not cut across what is known from microscopical research concerning the fertilization of the ovule by the pollen-grain. Here once again the greater can shed light upon the smaller and not only the other way.

73 § 58. The qualitative significance of the passage through the infinite or zero is brought out in all "degenerate" transformations, as when a finite spheroid flattens into a plane or contracts into a point (cf. Note 27 and the description of Plate IX, § 31). It is also a kind of transformation through the infinite when in the Cassini picture the curves which show a single sweep pass over *via* the lemniscate into the seemingly double forms. In all such cases the threshold form is different in character from those that emerge on either side, whether or no these show a qualitative difference from one another. A form of higher type dies and is reborn in passing such a threshold. By mathematical necessity the sequence of continuous transformation includes moments of sharp and sudden metamorphosis. Since this is evidently also Nature's way, the maxim *Natura non facit saltus* is incorrect, yet something of it remains true, for an ideal thread of continuity leads across the abysmal jumps.

Other remarkable examples occur in the study of continuous collineation-groups (Note 41). The invariant tetrahedron may degenerate in sundry ways by the fusion into one or two or more of its points and planes, bringing about a sudden and decisive change in the whole character of the resulting path-curves or lines of flow. Every such change is reflected in the number of times and in the ways in which zeros occur in the "canonical" matrices in which these different types of transformation find their simplest algebraic expression. (J. A. Todd, *Analytical and Projective Geometry*, 1947, Chapter V.) In the number and distribution of "functional infinitudes" there is a mathematical formative principle, qualitative rather than quantitative. This again seems to open out the way to a mathematical morphology of living forms and processes of growth.

It might perhaps be objected that in the calculus the transition from finite magnitudes to the infinitesimal has long been known to the mathematical physicist. But in working out such problems as the path of a projectile or the distribution of stresses in an elastic body the differential calculus is used, in this respect, only as a means to an end. The entities of which the movement or interrelation is imagined retain their finite magnitude. That they themselves should ever vanish or arise again from nothing is an idea to which the science of the twentieth century begins to awaken.

Differential geometry itself will of course undergo a change in emphasis if not in content when the planar aspect of space is treated as primary – on an equal level with the pointwise. In negative space it will be as natural to look by differentiation for the point of a curve, letting two tangent lines or three osculating planes melt into one, as in Euclidean space it is natural to start with the pointwise curve and look for the lines or planes. Mathematically this is obvious, but many well-known processes of Nature may appear in a new light when approached in this spirit.

74 § 58 E. F. Schumacher, author of *Small is Beautiful* (London 1979).

75 § 58 The work is being carried out at the Pfeiffer Foundation, Inc., Spring Valley, N.Y. 10977, U.S.A., founded by Dr. E. E. Pfeiffer (1899–1962).

76 § 58 A. John Wilkes directs the Flow Research and Sculpture Groups at Emerson College, Sussex, England.

INDEX

Repeated references to Rudolf Steiner and Goethe, and also to Modern Projective Geometry, occur throughout the book. These are omitted in the Index.

221